Agricultural Mechanization and the Evolution of Farming Systems in Sub-Saharan Africa

Prabhu Pingali
Yves Bigot
Hans P. Binswanger

Published for the World Bank
THE JOHNS HOPKINS UNIVERSITY PRESS
Baltimore and London

The Johns Hopkins University Press
Baltimore, Maryland 21211, U.S.A.

First printing February 1987

Library of Congress Cataloging-in-Publication Data

Pingali, Prabhu L., 1955-
Agricultural mechanization and the evolution of farming systems in Sub-Saharan Africa.

Bibliography: p.
Includes index.
1. Farm mechanization—Africa, Sub-Saharan.
2. Agricultural systems—Africa, Sub-Saharan.
3. Animal traction—Africa, Sub-Saharan.
4. Farm mechanization—Government policy—Africa, Sub-Saharan. I. Bigot, Yves. II. Binswanger, Hans P. III. Title.
S760.A357P56 1987 338.1'61'0967 86-27523
ISBN 0-8018-3502-X

Contents

Preface

This book owes important intellectual debts to Ester Boserup and to Hans Ruthenberg. In *Conditions of Agricultural Growth* Boserup sketches the evolutionary hypotheses which provided us with the framework for analyzing the links between population density, market access, and the evolution of farming systems from hand-hoe–based long fallow systems to permanent plow-based cultivation systems. Ruthenberg's seminal work, *Farming Systems of the Tropics,* traces this evolutionary pattern for many specific agroclimatic and soil environments with the exacting detail essential to understanding evolutionary trends in specific locations.

It was with this theoretical foundation that we tackled the central puzzle of this book: why is Sub-Saharan Africa not more mechanized? The answer cannot be found by applying the standard microeconomic framework of choice-of-techniques analysis, or even the framework extended to technical change, the induced innovation model. Because Sub-Saharan Africa has historically had abundant land, these frameworks imply that the region should be far more mechanized than it is today. To solve the puzzle and to provide the broader framework within which choice-of-techniques analysis and induced innovation provide additional insight, we relied on the evolutionary perspective of Boserup and Ruthenberg. Despite their contributions, the works of Boserup and Ruthenberg have occupied only the fringes of the consciousness of economists working on agricultural development.

We began our research by narrowly focusing on the mechanization question. We quickly ventured into carefully examining and testing grander evolutionary themes. But, following Ruthenberg's lead, we also tried to integrate the agricultural details into the evolutionary process to understand better the immense variations in mechanization trends which are observed across different agroclimatic regions, soil types, toposequences, and agricultural operations.

We obtained constructive reviews from Bruce F. Johnston and Graeme Donovan. Michael Lipton's extraordinarily detailed review deserves special mention: it challenged us to think more deeply about virtually every section of the manuscript. The book also benefitted from the comments of John McIntire, Paul Starkey, Raymond Naronha, and Bill Kinsey. We are also grateful to the many individuals who facilitated our fieldwork in the ten Sub-Saharan countries that we visited, in particular Willis Oluoch-Kosura, Mullegata Mekuria, Getachew Assamehew, and George Silumesi. Willis Oluoch-Kosura provided both guidance for our fieldwork in Kenya and a succinct account of the Kenyan mechanization situation. Statistical assistance was provided by Craig Lissner and Hans Tansen. Finally, we are indebted to David Howell Jones for editing the manuscript and to Gene Cummings and Rhonda Zaharna for typing and retyping it.

HANS P. BINSWANGER

Chapter 1

Introduction and Policy Overview

While the slow progress of both animal traction and motorization in Sub-Saharan Africa has been fairly well documented, we were motivated to undertake this study by the persistence of several puzzles. The following questions sparked our interest:

- Given that the rapid spread of mechanical equipment has historically been associated with an abundance of land, why has the process of mechanization been so slow and why is the hand hoe so tenacious in Sub-Saharan Africa? Why has the spread of mechanization in Sub-Saharan Africa been slower than in countries such as India, Pakistan, and China, where labor is abundant and wages are low?
- Why have animal traction and tractorization spread fairly rapidly in restricted pockets of Africa, but left neighboring agroclimatic regions untouched? Small farmers in Sukumaland, Tanzania, for instance, use ox plows for cultivating valley bottoms, while the surrounding upland areas are cultivated with hand hoes.
- Why have some cattle-owning farm households historically failed to use their own oxen in cultivation?
- Why have attempts of governments and donor agencies to bypass the animal-traction stage for direct tractorization through tractor projects repeatedly failed?

Our primary objective was to identify the conditions that lead to the transition from hand tools to animal traction and the further transition to the tractor. Early in the study, however, we found that any investigation of agricultural mechanization independent of the context of the farming systems would be incomplete, because the transition from the hand hoe to the plow is closely associated with the intensification of the farming system, a point emphasized in the work of both Ester Boserup and Hans Ruthenberg.

We therefore assumed the following objectives:

- to identify more clearly the conditions under which societies evolve from shifting cultivation to permanent cultivation of land
- to analyze the way agroclimatic, soil, and infrastructural constraints accelerate or retard this evolution of farming systems
- to specify the stage in the evolution of farming systems at which the conditions are most appropriate for the transition from the hand hoe to the plow, whether animal-drawn or tractor-drawn
- to explain why despite repeated extension efforts some permanent farming systems persist in the use of the hand hoe
- to evaluate the policy-planning and project framework within which to promote the future growth of agricultural mechanization, whether based on animal traction or mechanical power.

Methodology

This study is based on an analytical review of the literature on agricultural mechanization in Sub-Saharan Africa. The literature, however, has many gaps and omissions. It was therefore complemented by field visits to fifty locations in ten countries of Sub-Saharan Africa. In the review of literature in English and French historical and agroclimatic information, as well as information of an economic nature, was emphasized. The review provided us with background information on countries such as Ethiopia and Botswana, in which animal traction had long been established; on regions such as Tanzania and central Burkina Faso, in which it began to spread after World War II; and on regions such as central Côte d'Ivoire and southern Nigeria, where it was tried but failed to have any effect.

Extensive field visits were made to the regions so identified. They are located on map 1-1. (Table 1-1, at the end of the chapter, provides summary information on each location. At each location systematic group interviews were conducted to elicit information on farming systems, tools used, and technical change.)

The conclusions that follow have been derived in a variety of ways. Some are well established in the literature, which is simply summarized here. Others are derived by synthesizing and analyzing data from existing studies. A third set of conclusions emerges from the analysis of our own data generated in the field visits. Finally, some conclusions emerge

Map 1-1. *Places Visited in Sub-Saharan Africa*

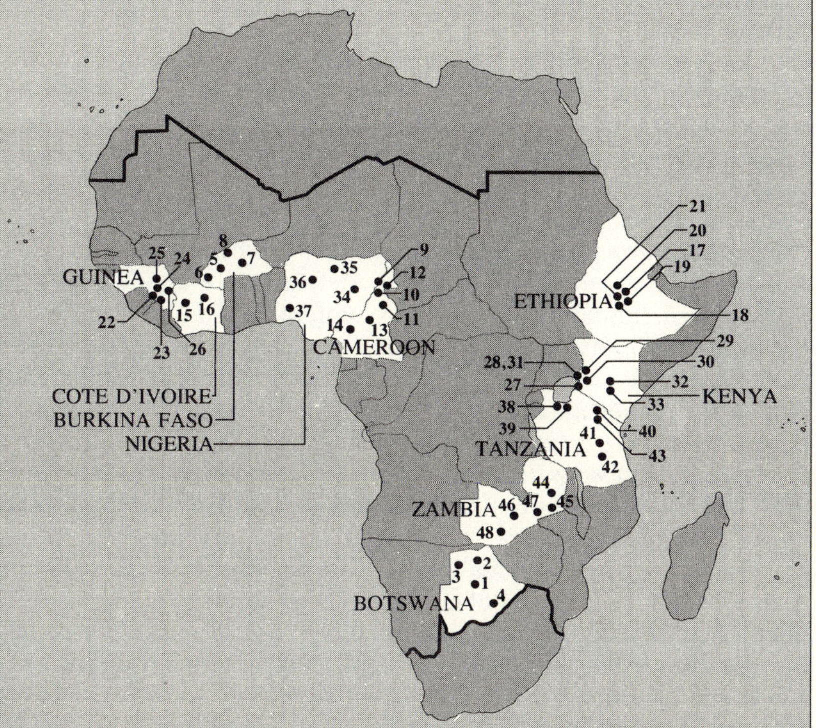

Note: The numbers refer to the list of locations visited given in table 1-1.

as logical implications of earlier conclusions. The basis for a conclusion will in general be clear in the chapter in which it is presented.

Conclusions and Policy Implications

The primary finding of this study is that the solutions generated by African farmers for increasing food production from a given area of land have been strikingly similar to the solutions found historically in other parts of the developed or developing world. African farmers have responded to increasing population density or increased demand for agricultural output with an expansion of the area cultivated, larger investments in land, and innovations in mechanical technology and manuring systems.

Population growth and access to markets are the main determinants of agricultural intensification, which is defined here as the movement from forest-fallow and bush-fallow systems of cultivation to annual and multiple-crop systems, whereby plots of land are cultivated one or more times a year. The existence of a positive relation between population density and agricultural intensification has been described by Boserup. An increase in population density or the amount of land available per person causes a reduction in fallow periods—that is, the movement to more permanent cultivation of land. In addition to verifying Boserup's hypothesis, we also find that improvements in access to markets through better roads and transport facilities have a similar positive effect on the intensity of land use. Better access to markets leads to intensification for two reasons: First, higher prices and elastic demand for tradable goods mean greater marginal rewards for effort, so farmers will begin to cultivate larger areas. Second, higher rewards to labor encourage immigration into the area from neighboring regions where transport costs are higher.

We have not encountered any sparsely populated areas under annual cultivation or very densely populated areas under forest-fallow or bush-fallow systems.

Heavier soils come to be used when population density rises or access to markets improves. Deep soils that have a high clay content have a greater capacity to hold water and nutrients. For that reason they typically provide a greater and more certain yield than lighter soils, but they are often difficult or impossible to cultivate in the absence of investments in water control or drainage. When population density is low, therefore, and neither animal nor motor power is available, lighter soils with a low clay content are preferred over the deep, clayey soils. As population density increases the labor supply increases, making it possible to undertake the investments in labor for irrigation and drainage. The heavier soils become intensively cultivated because they offer higher marginal returns on investments in labor, fertilizer, and land, especially for rice cultivation. Population pressure therefore leads to a reversal in the preference for one type of land over another, the deeper, heavier soils being preferred over the lighter more easily worked soils.

Agricultural intensification leads to an increase in yield per hectare. This conclusion emerges clearly from an analysis of existing studies. The higher yields are obtained because farmers respond to declining availability and fertility of land by increasing the input of labor to improve land preparation, crop husbandry, and soil fertility, and because farmers move to heavier soils, which are more responsive to labor and other inputs. In the absence of animal or motor power, however, the added labor requirements typically lead to a decline in yield per man-hour.

In the absence of the plow agricultural intensification leads to an increase in agricultural employment. Holding mechanical technology constant, total labor input per hectare is positively correlated with intensity of farming. Ruthenberg, for example, showed that the movement from growing rice under forest fallow to growing it under annual cultivation is associated with an increase in total labor input per hectare from 770 hours in Liberia to 3,300 hours in Cameroon. This increase arises because of an increase in both the number of field operations and the intensity with which they are performed. More labor is also required for investments in irrigation, drainage, leveling, and terracing.

The use of organic fertilizer is positively associated with the scarcity of land. In land-abundant areas the fertility of the soil is maintained by periodic fallowing of land. As the fallow periods become shorter the use of organic fertilizer begins to emerge. At first these fertilization techniques are rudimentary, often involving no more than a periodic transport of household refuse to the fields. As farming intensifies, more highly evolved manuring and composting techniques are used. Under extreme scarcity of land, crop rotation, including legumes or green manures, emerges. An alternative is to increase the use of chemical fertilizers as a substitute for labor-intensive manuring techniques. General use of chemical fertilizers is still rare in Sub-Saharan Africa, however, although the use of fertilizers for selected cash crops such as cotton and groundnut is becoming increasingly common.

The transition from the hand hoe to the animal-drawn plow is only profitable at higher intensities of farming. The big puzzle is why an apparent labor-saving technology, the animal-drawn plow, is introduced only at higher population density—that is, when the number of laborers per unit of area has already increased. The puzzle has the following solution: When farming intensity increases beyond the forest-fallow stage, both the number and the intensity of farm operations increase, and the labor required per unit of output—and per unit of land—rises faster than the availability of labor. Animal traction is indeed labor-saving with respect to the high requirement of labor per unit of output of hand-hoe cultivation in the intensive farming systems. It does not save on the relatively low requirement of labor per unit of output of hand-hoe farming under forest-fallow or bush-fallow, however, but increases it. It does not improve the welfare of the forest-fallow or bush-fallow farmers, who therefore reject it.

The transition from the digging stick and the hand hoe to the animal-drawn plow is therefore closely correlated with the evolution of the farming system and cannot be understood by using a simple choice-of-techniques analysis familiar to economists. This transition is not cost-effective in forest-fallow and bush-fallow systems because fields must not only be cleared but also destumped if the plow is to be used effectively. This is an arduous task when the density of trees is still high. Shorter fallows lead to lesser density of trees and roots, and the cost of destumping declines. Root density is minimal by the time the grass-fallow stage is reached, and it is here that animal power becomes the economically dominant technology. By that time, moreover, trypanosomiasis, the main animal disease problem of forest-fallow systems, becomes less of a constraint on ownership and use of animals. Finally more grassy vegetation provides more abundant fodder.

The exact level of intensity at which the switch to the plow improves welfare varies greatly according to local conditions. It occurs earlier where heavy soils make hand-hoe cultivation particularly arduous, for example, or where a longer growing season allows for high capacity utilization of the plow and its source of power, or where a well-developed market for meat makes the raising of livestock more attractive.

The substitution of mechanical tillage for hand tillage generally has a minimal effect on yields. Experimental studies show that the yield effects of tillage vary with type of soil and crop. They are the highest on heavy soils, such as black clays, and least on light sandy soils. On a given type of soil the yield benefits of tillage are greatest for rice and least for groundnut. Yet objective comparisons of hand-hoe farms with animal-traction farms have consistently failed to show any significant yield effects. This is consistent with the worldwide finding that the switch from animal traction to the tractor has little or no effect on yields. Experimental evidence shows that it is often possible to increase yields by improving the quality of tillage. If farmers fail to achieve increases in yields it implies that they do not use new sources of power to improve the quality of tillage. The switch in sources of power is instead motivated by a combination of other benefits such as expansion of the area under cultivation and labor savings in tillage, weeding, and transport.

It is not unusual for farmers to persist in the use of the hand hoe in some intensively cultivated regions. Mountainous tropical highland areas are a primary example of intensively cultivated areas in which the use of the hand hoe persists, for two reasons: the steep terrain is often a constraint to the use of animal or tractor power, and many tropical highland areas have a comparative advantage in the production of tree crops or milk rather than field crops, so opportunities for the use of mechanical tillage equipment are limited. Although small quantities of field crops are grown for subsistence, the plots are small and the season is long, so the hand hoe continues to be the dominant technology.

It has not been possible to use the tractor to accelerate the evolution of farming systems to permanent cultivation significantly. This a consequence of the conclusion that the plow becomes appropriate only at the bush-fallow or early grass-fallow stage. All tractor projects we have analyzed in late bush-fallow or early grass-fallow systems have failed.

In permanent cultivation systems, an opportunity sometimes exists for replacing draft animals with tractors. Whether this transition is cost-effective must be analyzed case by case. While the transition from the hand hoe to animal-draft power is closely tied to the evolution of farming systems, the transition from animal-draft power to the tractor falls within the realm of the choice-of-techniques analysis. Consider a society with a well-established farming system based on the animal-drawn plow. With the advent of the tractor into the region the farmer can choose between two sources of power for performing various operations, such as tillage and transport, while holding all other aspects of the farming system constant. Whether he chooses animal draft or the tractor for a particular operation depends on the relative costs of labor and capital, the interest rate, his potential utilization of capacity, the size of his farm, the availability of fodder, the relative maintenance costs of animals and tractors, and the difficulty of obtaining spare parts, fuel, and repair services. It should be emphasized that the choice-of-techniques analysis implies that not all operations are transferred to the new source of power and that a coexistence and a complementarity of power sources can be seen for prolonged periods.

The direct transition from the hand hoe to the tractor at the late bush-fallow and early grass-fallow stages is not cost-effective. In addition to the choice-of-techniques issues already discussed, a farmer at the late bush-fallow or early grass-fallow stage who is contemplating a choice between animal-drawn and tractor-drawn plows needs to evaluate several conditions. These include the costs of destumping for animal and tractor tillage, the costs of animal maintenance, considering the availability of fodder and the risks of animal disease, and the cost of losing tractor-animal complementarity.

Destumping costs are higher for tractor farmers because a much higher quality of destumping is required in order to minimize the damage to tractor-drawn implements. Farmers who use animal draft, on the other hand, can work around the stumps, so animal-drawn plows can be adopted even in short-fallow systems calling for three or four years of fallow.

By the late bush-fallow and early grass-fallow stages, the supply of animals in the area increases because of the ease of obtaining fodder and a reduction in the incidence of trypanosomiasis. Animal maintenance costs are therefore lower than in the long-fallow systems, and the farmer can benefit from various types of interaction between crop and livestock, such as the production of manure. By choosing not to use animals for draft the farmer would forgo an additional important benefit that he can derive from incorporating animals into his farming system. By going directly to the tractor, moreover, the farmer would also lose the complementarity between tractor and animals, which is particularly valuable where wages are low.

There are only two environments in which the direct transition to tractors may be cost-effective. These are valley bottoms and flood plains, where irrigated or flooded rice is cultivated, and the grassy savannas in the semiarid zones. The natural vegetation in both kinds of land does not include bushes or trees but is primarily a grass cover, so the constraint of high destumping costs is not a problem, even when population density is low. Further case-by-case analysis is required, however, to determine whether animal-drawn or tractor-drawn plows are more cost-effective in a given location. The choice between the two is based on the choice-of-techniques question, learning costs, and the costs of infrastructure.

The simultaneous transfer of all agricultural operations to a new source of power, animal or motor, is usually not economically attractive.

Agricultural operations can be grouped according to the intensity with which they require power—or energy—in relation to the control functions of human judgment. Operations such as land preparation, transport, milling, grinding, and threshing are power-intensive, while weeding, sifting, winnowing, and fruit harvesting are control-intensive. New sources of power are always used first for power-intensive operations. It is for these that the use of the new source of power has the greatest comparative advantage and may be profitable even when wages are low. The transfer of control-intensive operations to the new source of power becomes profitable only when wages are high or rising. In land-scarce economies, where the nonagricultural demand for labor is low, operations such as weeding, interculture, and harvesting continue to be performed by human or animal power. This tractor-animal complementarity ought to be recognized and encouraged in projects that promote mechanization.

Historically, too much emphasis has been placed on a self-defeating effort to tie biological technology to mechanical technology. The tying of "improved" crop-husbandry practices, such as the incorporation of crop residues into the soil, to projects that promote animal traction has typically been unsuccessful. The reason is again that labor saving and area expansion are the main motivating forces in the adoption of animal traction, while most husbandry practices require an extra input of labor. More broadly, our data for Africa and the preexisting literature imply that high-yielding varieties and fertilizers are no precondition for mechanization, nor is mechanization—except, perhaps, the mechanization of irrigation—a precondition for the adoption of high-yielding varieties and fertilizers. The history of North America and other land-abundant areas shows that where the farming systems have allowed or required mechanization, it has often preceded by decades the adoption of any biological technology. In land-scarce countries, such as Japan, however, biological technical change occurred toward the early part of the twentieth century, while the widespread use of mechanical technology was a more recent phenomenon. The direction of technical change—that is, the emphasis on mechanical or biological technology—is determined by the factor endowments of an economy, and attempts to encourage both concurrently as equal parts of a technical package have generally failed.

The profitability of animal-drawn or tractor-drawn equipment is constrained by the length of the growing season. The length of the land-preparation period is an important determinant of the capacity utilization of animal-drawn and tractor-drawn plows. The shorter the period of land preparation, the lower the rates of utilization. Accordingly, the shorter the season, the higher the costs per unit of area that is to be cultivated. The period during which mechanical tillage equipment can be used in the arid and semiarid tropics is extremely short. Timeliness of plowing is crucial if there is to be adequate time for a crop to emerge before soil-moisture stress becomes a problem. Because of the time-bound nature of plowing in the dry tropics, timing conflicts of potential users make it difficult to establish markets for the rental of equipment. For that reason equipment costs cannot be spread over a large area. Plowing periods are longer in the subhumid and humid tropics, in areas where the rainfall regime is bimodal, and in high altitudes. Rental markets are easier to establish and develop in these areas, and the cost of equipment can be spread among several users. Utilization of tractor capacity can be increased by providing contract-hire operations that span several rainfall zones.

Within any given area, animal draft is typically used first by large households, and tractors are first introduced on larger landholdings. Large households typically find it easier than do smaller ones to use draft animals to full capacity, even if at first the area cultivated per worker in the large household is not larger than that in the smaller household. Large households usually find it easier, moreover, to finance the acquisition of equipment and animals. Animal draft then allows these households to increase the area cultivated per worker. Where land is still fairly abundant, this pattern tends to be followed at the expense of fallow rather than at the expense of other households.

Where tractors replace draft animals they do so first on the largest farms. Since this transition typically occurs when land is already fairly scarce, the area expansion that follows tractorization is often at the expense of other households, rather than representing a net addition to cultivated areas.

Lack of animal-husbandry or mechanical skills is only a short-run constraint to the use of animal or tractor power. In our study we found that where farming systems were conducive to use of the plow, African farmers readily acquired the skills needed to operate animal-draft or tractor equipment. We found, moreover, that many of the skills required for the efficient use of animal-draft or tractor equipment had been acquired by these societies before they had acquired this equipment. A majority of the tribes in Sub-Saharan Africa outside the humid forest zone, for example, are crop cultivators who keep cattle for milk or meat and as a store of wealth. These farmers already possess animal-husbandry skills, although they may lack the ability to train their animals for draft purposes. Even in the humid forest zone, where trypanosomiasis precludes the keeping of livestock, farmers could and did acquire livestock-husbandry skills once it was possible and profitable for them to do so. Similarly, the use of bicycles, motorcycles, and mechanical mills has become common all over Sub-Saharan Africa, even in areas where animal traction and tractors are not used. The mechanical skill required for the use of this equipment is as complex as that required for the use of animal-drawn implements. We also found that where animal-draft or tractor equipment was in use, workshops capable of servicing this equipment emerged fairly rapidly and were located where they were accessible to the users.

The planning of project interventions to promote any form of mechanization is best done at the regional or subregional level rather than at the national level. The profitability of mechanical tillage varies with farming system, soil type, and agroclimatic zone. The potential for use of the plow is greater on more intensively cultivated areas (grass-fallow systems and above), in heavier soils such as vertisols, and in areas in which growing periods are longer. In these variables, regions and subregions of all but very small countries vary enormously. The profitability of using the plow therefore varies widely among locations, and our analysis so far indicates that the selection of an appropriate location is the single most important determinant of the eventual success of a project.

Identifying and alleviating short-run constraints in the transition to animal power is the most appropriate goal of government intervention. The profitability of using the plow is dependent on the farming system, soil type, and agroclimatic conditions. Mechanization policy and interventions can, at best, affect these factors marginally. At appropriate locations, however, mechanization interventions can alleviate some of the short-run constraints to the transition to the plow. The primary short-run constraints are:

- The availability of credit. Credit is more difficult to provide for animal power than for tractors because the loans are smaller and administrative costs are therefore high, and repossession of animals in the case of default is more difficult than repossession of tractors. Private channels of credit are therefore not well established, and a case can be made for a government program as part of an extension effort to promote the use of animal traction.
- Veterinary services. Most veterinary and livestock services in Sub-Saharan Africa have monitoring and immunization programs in an attempt to control diseases such as rinderpest and

foot-and-mouth disease. In addition to such programs, draft animals require specific treatment for injury and special feeding designed to improve their strength and tractive power. It is often not easy to induce the livestock ministry—or the government—to give special emphasis to these problems. To mobilize resources for providing specific veterinary care for draft animals requires a coordinated program between the ministries of agriculture and livestock.

- Training programs for animals and farmers. During the initial stages of the use of animal traction, government-sponsored training programs reduce the learning costs substantially. These training programs are best organized in small regional centers under conditions similar to those in the local farm environment. The training centers can also be used for training farmers in the use of new equipment. There is not usually a necessary link between training farmers to use draft power, however, and the diffusion of other technologies. Associating nonessential new techniques with ox-training activities may cause programs to grow to unmanageable size and may impede the transfer of basic animal-traction skills.
- Blacksmith training. The widespread use of animal-draft equipment will create a demand for maintenance and spare-parts services. Experience in Sub-Saharan Africa has shown that the establishment of central workshops to which farmers must come to have equipment repaired or to procure spare parts is not sufficient. Village blacksmiths, on the other hand, have been more effective in providing repair services. Well-trained blacksmiths can also be useful auxiliaries in establishing plants for the assembly or manufacture of equipment. The training and retraining of blacksmiths is therefore a useful component of animal-traction projects.

Transition from animal-draft power to the tractor, where it is appropriate, occurs rapidly and does not depend on government intervention. Where the transition to the tractor is profitable, rapid adoption of tractor power often takes place with minimal government involvement. Supply-side constraints are not a significant bottleneck in their adoption. Repair and service facilities follow demand—where they are not restricted by government policy—and privately operated machines typically have long lives and high rates of utilization.

The historical record is quite clear on government intervention. Tractor mechanization in the countries that belong to the Organisation for Economic Co-operation and Development (OECD) did not depend on direct government intervention in machinery development, production, choice of technology, or finance. The most successful experiences in the developing world, such as the mechanization of milling, pumping, or harvest processing, did not depend on special interventions either. Once economic conditions have led to an effective demand for machinery, private firms in the developed world have responded rapidly. The response of the private sector has been equally rapid in developing economies as diverse as Thailand; India; Taiwan; Mexico; Odienné, Côte d'Ivoire; and Arusha province, Tanzania.

Government tractor-hire services for primary tillage operations have not been economically viable. Seager and Fieldson assembled a large number of examples of government tractor-hire operations. In twenty of the twenty-one cases that they examined government tractor-hire services for tillage failed. Several causes for their failure can be enumerated: There are typically no economies of scale associated with operating large tractor fleets. These services are generally characterized by severe problems of motivating their staffs to high levels of efficiency; tractor drivers typically do not or cannot be made to benefit financially from working long hours and maximizing the number of hectares tilled or from keeping tractors in good condition during the peak period. And the rate of utilization of the equipment is very low, since the demand for the service is mainly for tillage. Unlike private contractors, operators of public hire schemes generally find it difficult to allow tractors to cross agroclimatic zones to increase rates of utilization.

The one exception found by Seager and Fieldson was a bulldozer rental service in Pakistan organized for clearing and leveling land. Land clearing and leveling operations are not time-bound, so the service can achieve a high rate of utilization and can survive economically.

The characteristics of successful tractor-rental operations are (1) private ownership by farmers or small entrepreneurs; (2) concentration on power-intensive operations; (3) where plowing seasons are short, solution of the capacity-utilization problem through migration of tractors among agroclimatic zones; and (4) continued use of animal traction for less power-intensive operations.

Where private rental markets fail because conditions 2, 3, and 4 cannot be met, public hire schemes will not work either, and where private rental arrangements are feasible, public hire schemes are not needed.

In the promotion of tractor use government policy should encourage the market test. Markets are capable of sorting out the appropriate locations for tractorization and the speed with which it should proceed. Government involvement is best limited to ensuring dissemination of market signals, which can best be achieved in an environment where factor price distortions are minimal and there are a minimum of quantitative restrictions of foreign-exchange bottlenecks to impede the import of a diverse spectrum of machines, spare parts, and fuels.

The following interventions have frequently accelerated tractorization beyond the pace that was economically justified and at which it could add to employment problems and other social problems:

- *Preferential tariff treatment for tractors.* If there is a general revenue tariff, there is no reason to exempt tractors or spare parts.
- *Preferential credit terms.* Tractor loans are preferred by rural credit institutions because they are large chunky loans to wealthy farmers who can provide collateral and because tractors can serve as their own collateral. There is therefore no need to encourage further tractorization with special, often subsidized, allocation of credit. Such loans may contribute to the emergence of a class of large farmers, moreover, while further marginalizing small ones.
- *Minimum-wage laws.* Such laws are often used as incomes policies, and the wages are generally too high in relation to the market wage rate. They are only effective in the organized plantation sector, where the immediate response is to substitute capital-intensive equipment for labor. The net effect is often to reduce the income of labor in this sector.

Private initiative has been the dominant force in the generation of mechanical innovations and the development of an agricultural machinery industry. Mechanical technology is sensitive to agroclimatic factors such as soils, terrain, and rainfall regimes and to economic factors such as wage rates, the availability of capital, farm size, and the materials that are available. Where there is a divergence in either environmental or economic conditions, opportunity for the direct transfer of mechanical technology is limited. Accordingly, where factor endowments warrant it and where the benefits from innovation can be appropriated, a great deal of inventiveness or adaptation of mechanical technology to meet local conditions can be seen. In the early phases of mechanization such work is usually done by small manufacturers or workshops in close association with farmers. This process provides direct solutions by mechanically minded individuals to problems perceived by farmers. In 1880, for instance, there were 800 distinct models of plow advertised for sale in the United States. Early successful machinery innovations in the developing world reveal similar reliance on small workshops and direct contact with farmers. The emergence of a diversified machinery industry out of small shops in the Indian Punjab and the power-tiller industry in Thailand and the Philippines all followed similar patterns. In the early phases small workshops have a distinct advantage over large corporations because of the location-specific nature of the innovations and the ability of the producers to gain from their innovative efforts through sales. The contribution of large corporations increases with the passage of time but continues to be most important in engineering optimization.

Given this dominance of individual initiative in the development and adaptation of agricultural machinery, the appropriate government interventions to encourage mechanical innovation are patent laws and other legal provisions that enable innovators to capture a portion of the economic rent generated by their innovations; testing and the dissemination of information; and support of agricultural engineering education. Finally, it should be noted that efforts to protect the domestic agricultural machinery industry through import controls have often been self-defeating. This is because the small innovators no longer have access to models or a wide range of engines and other parts to design machines adapted to local conditions.

High output prices accelerate the pace of intensification and mechanization, provided that they are transmitted to the farm gate. Price incentives motivate agricultural intensification and the associated farm investments, including investments in mechanical equipment. Prices, however, are not sufficient encouragement for technical change. Among the other requirements are the existence of adequate market infrastructure. High border prices cannot be transmitted to the farm level if interregional roads are deteriorating and if there are no farm-to-market roads. In countries in which population density is low, however, the per capita cost of infrastructural development is high.

We hope that this book will be useful to three audiences. For students of agricultural development who will be interested primarily in part one, it provides tests of the theories of the evolution of farming systems and agricultural technology associated with Ester Boserup and Hans Ruthenberg. The second audience consists of specialists in agricultural mechanization, whether they be engineers or social scientists. For them the book provides a comprehensive treatment of mechanization issues in Sub-Saharan Africa. In particular it deals at length with the problems associated with the transition from the hand hoe to more advanced forms of mechanization, a topic of great importance for Africa but on which much less systematic work has been done than on issues of replacing animal draft with motorized sources of power. The third audience is the development practitioners who must design policy and project interventions in Sub-Saharan Africa. For them the book also contains a number of policy and project prescriptions. Chapter 8, on the design of animal-traction projects, is addressed almost exclusively to this audience.

Table 1-1. *Summary Information concerning Locations Visited*

Country, region, and observation number	Annual rainfall (millimeters)	Population density[a]	Access to markets	Farming system	Tools used	Primary crop	Secondary crop
Botswana							
1. Palo Sehtla district	540	n.a.	Good	Short fallow	Animal traction[b]	Sorghum	Maize
2. Eastern Ngamiland	510	n.a.	Very poor	Short fallow	Animal traction	Sorghum	Maize
3. Western Ngamiland	510	n.a.	Very poor	Short fallow	Animal traction	Millet	Sorghum
4. Gaborone	375	n.a.	Good	Short fallow	Animal traction[b]	Maize	Sorghum
Burkina Faso							
5. Boromo region	950	40	Excellent	Short fallow	Animal traction	Sorghum	Millet
6. Volta Noire	1,050	35	Good	Short fallow	Animal traction	Sorghum	Maize
7. Ouagadougou	550	80	Fair	Emerging annual cultivation	Animal traction	Millet	Sorghum
8. Yatenga	850	50	Very poor	Emerging annual cultivation	Hoe	Millet	Groundnut
Cameroon							
9. Mokolo	1,000	80	Fair	Short fallow	Hoe	Cotton	Sorghum
10. Guider	850	40	Fair	Short fallow	Animal traction	Cotton	Sorghum
11. Tupairi	950	80	Fair	Short fallow	Animal traction	Cotton	Sorghum
12. Maroue	700	100	Fair	Annual cultivation	Animal traction	Cotton	Sorghum
13. Guatala	600	40	Poor	Short fallow	Animal traction	Cotton	Sorghum
14. Bamenda	2,500	40	Excellent	Short fallow	Hoe	Maize	Yam
Côte d'Ivoire							
15. Morondo	1,350	5	Poor	Forest fallow	Hoe	Rice	Groundnut
16. Northern region	1,250	15	Good	Bush fallow	Animal traction[b]	Maize	Cotton

(Table continues on the following page.)

Table 1-1 *(continued)*

Country, region, and observation number	*Annual rainfall (millimeters)*	*Population density*[a]	*Access to markets*	*Farming system*	*Tools used*	*Primary crop*	*Secondary crop*
Ethiopia							
17. Debra Behram	800	38	Good	Emerging annual cultivation	Animal traction	Barley	Horsebean
18. Ada Werada region	845	73	Excellent	Annual cultivation	Animal traction	Wheat	Maize
19. Nazreth area	745	48	Excellent	Annual cultivation	Animal traction	Maize	Sorghum
20. Chilalo Awraja	900	56	Good	Annual cultivation	Animal traction	Wheat	Maize
21. Shoa region	1,067	62	Poor	Emerging annual cultivation	Animal traction	Maize	Other
Guinea							
22. Guéckédou, village 1	2,400	40	Poor	Bush fallow	Hoe	Rice	Other
23. Guéckédou, village 2	2,300	40	Poor	Bush fallow	Hoe	Rice	Other
24. Kissi Dougou	2,000	10	Very poor	Forest fallow	Hoe	Rice	Cassava
25. Kankan	1,600	20	Very poor	Bush fallow	Animal traction	Groundnut	Rice
26. Higlit Guinea	1,300	5	Very poor	Bush fallow	Animal traction[b]	Maize	Rice
Kenya							
27. South Nyanza	1,000	114	Good	Annual cultivation	Animal traction	Sugarcane	Other
28. Busia	1,250	140	Good	Emerging annual cultivation	Animal traction[b]	Maize	Cassava
29. Bungoma	1,250	290	Good	Annual cultivation	Animal traction[b]	Other	Maize

30. Kakamega	1,250	294	Good	Annual cultivation	Animal traction[b]	Maize	Bean
31. Siaya	1,150	186	Good	Annual cultivation	Animal traction	Maize	Other
32. Embu	750	50	Good	Short fallow	Hoe[b]	Bean	Maize
33. Machakos	750	50	Good	Emerging annual cultivation	Animal traction	Other	Bean
Nigeria							
34. Bauchi state	750	56	Good	Emerging annual cultivation	Animal traction	Millet	Sorghum
35. Kano state	850	120	Good	Annual cultivation	Hoe	Maize	Millet
36. Kaduna state	1,100	75	Excellent	Annual cultivation	Animal traction[b]	Sorghum	Cotton
37. Oyo state	1,350	75	Fair	Short fallow	Hoe	Yam	Maize
Tanzania							
38. Mwanza hills, Kimba district	1,083	93	Good	Annual cultivation	Hoe	Cotton	Cassava
39. Mwanza valley, Kimba district	1,083	93	Good	Annual cultivation	Animal traction[b]	Rice	Sorghum
40. Arusha, Arumeru district	927	82	Excellent	Annual cultivation	Animal traction[b]	Maize	Bean
41. Northern Kilosa	984	11	Poor	Bush fallow	Hoe	Maize	Pulse
42. Southern Kilosa	1,036	11	Poor	Bush fallow	Hoe	Maize	Pulse
43. Arusha district	927	82	Excellent	Annual cultivation	Animal traction[b]	Maize	Bean

(Table continues on the following page.)

Table 1-1 *(continued)*

Country, region, and observation number	*Annual rainfall (millimeters)*	*Population density*[a]	*Access to markets*	*Farming system*	*Tools used*	*Primary crop*	*Secondary crop*
Zambia							
44. Chipata	1,014	14	Very poor	Forest fallow	Hoe	Cotton	Sorghum
45. Katete	1,000	8	Good	Short fallow	Animal traction	Maize	Groundnut
46. Kandungo	844	4	Fair	Bush fallow	Hoe	Maize	Sorghum
47. Marushi	844	4	Good	Bush fallow	Hoe	Maize	Groundnut
48. Mazabuka	1,065	18	Excellent	Annual cultivation	Animal traction[b]	Maize	Cotton
India							
Dokur, Andhra Pradesh	928	245	Excellent	Multiple cropping	Animal traction[b]	Rice	Groundnut
Aurepalle, Andhra Pradesh	681	268	Good	Multiple cropping	Animal traction	Rice	Castorseed

Rampura, Madhya Pradesh	1,353	94	Fair	Multiple cropping	Animal traction	Wheat	Lentil
Papda, Madhya Pradesh	1,353	94	Good	Multiple cropping	Animal traction[b]	Wheat	Lentil
Shirapur, Maharashtra	597	84	Good	Multiple cropping	Animal traction	Sorghum	Pearl millet
Kalman, Maharashtra	569	65	Excellent	Multiple cropping	Animal traction	Sorghum	Pearl millet
Kanzara, Maharashtra	818	156	Good	Multiple cropping	Animal traction	Cotton	Sorghum
Kinkeda, Maharashtra	818	133	Excellent	Multiple cropping	Animal traction	Cotton	Sorghum

n.a. Not available.

Note: Eight observations made in India have been included for comparison with the African observations.

a. Persons per square kilometer.

b. A few tractors were in the area.

Part One

The Decision to Plow

Chapter 2

Farming Intensity and the Plow: The Analytical Framework

In most agricultural environments the transition from the hand hoe to the plow, whether animal-drawn or tractor-drawn, is closely associated with the evolution in farming systems from forest-fallow to annual cultivation. Under forest-fallow systems hand-hoe cultivation is capable of producing output at the lowest cost per unit. Land-preparation and weeding requirements are minimal, and cleared land does not have to be destumped before planting. To use the plow with this system would require an increase in the number of operations performed and therefore an increase in the total amount of labor required. As the length of fallow decreases, however, cultivation and weeding requirements increase, while the costs of destumping decrease. At the short-fallow stage, therefore, it becomes economically attractive to switch to the plow.

The main purpose of this chapter is to explain why an apparently labor-saving technology, the animal-drawn or tractor-drawn plow, is not used in forest- and bush-fallow systems when population densities are low and labor is scarce and why it becomes attractive only when farming intensities increase, usually as a consequence of population growth. The answer is that in the shift to annual cultivation the demand for labor per unit of output increases faster than the supply of labor. This is because the number of field operations and the intensity with which they have to be performed increases rapidly as the length of fallow is shortened. It is only in relation to the labor requirements of the short-fallow and permanent systems of cultivation that the plow is labor-saving, not with respect to labor requirements of forest-fallow or bush-fallow systems.

The exceptions to the sequence just discussed and the influence of trypanosomiasis in slowing the emergence of animal traction will also be discussed.

The Evolution of Farming Systems in the Tropics

The evolutionary movement from shifting cultivation from one plot of land to another after two or three years to the permanent cultivation of plots of land involves an increase in the frequency with which plots of land are cultivated.[1] Consistent with Boserup (1965) and Ruthenberg (1980), we define this evolution as agricultural intensification. Again following Ruthenberg, we measure the level of intensity by the frequency with which land is cultivated. Agricultural intensification is induced by increased demand for cultivated land. The demand for cultivated land can rise because of population growth, increased market demand for agricultural products, or reduced transportation costs. A detailed discussion of these and other determinants of agricultural intensification is deferred until chapter 3, in which some empirical evidence on the evolution of farming systems will be found.

The consequences of agricultural intensification are a transition from the hand hoe to the plow; an increase in investments for destumping, drainage, and terracing; an increase in the maintenance of soil fertility through manuring; and the movement from general use rights to specific land rights. In this chapter we shall be particularly concerned with the transition from the hand hoe to the plow. In order to understand this transition, however, it is necessary to examine some of the concurrent changes, specifically changes in the number of farming operations performed and—for given technologies—the consequent increase in the labor required per unit of output.

The existence of a positive correlation between the intensity of land use and population density has been shown by Boserup (1965, 1981). During the neolithic period forests covered a much larger part of the land than today. The replacement of forests by bush and grassland was caused by, among other things, a reduction in fallow periods associated with increasing population density. According to Boserup, "The invasion of forest and bush by grass is more likely to happen when an increasing population of long fallow cultivators cultivate the land at more and more frequent intervals" (1965, 20).

Table 2-1—in a stylized way—presents the relation between population density and the intensity of the agricultural system. At sparse population density—no more than four persons a square kilometer—depending, of course, on the fertility of the soil, the prevailing form of farming is the forest-fallow system. A plot of forest land is cleared and cultivated for one or two years, then allowed to lie fallow for twenty to twenty-five years. In most climatic zones this period of fallow is sufficient to allow regrowth of the forest. An increase in population density

Table 2-1. *Food-Supply Systems in the Tropics*

System[a]	*Farming intensity (R value)*[b]	*Density of population*[c] *(persons per square kilometer)*	*Tools used*[d]
Gathering (G)	0	0–4	None
Forest-fallow (FF)	0–10	0–4	Axe, machete, and digging stick
Bush-fallow (BF)	10–40	4–64	Axe, machete, digging stick, and hoe
Short-fallow (SF)	40–80	16–64	Hoe, animal traction
Annual cultivation (AC)	80–120	64–256	Animal traction and tractor

a. Description of food-supply systems:
Gathering—wild plants, roots, fruits, nuts
Forest-fallow—one or two crops followed by fifteen to twenty years of fallow
Bush-fallow—two or more crops followed by eight to ten years of fallow
Short-fallow—one or two crops followed by one or two years of fallow; also known as *grass-fallow*
Annual cultivation—one crop each year
Multiple cropping—two or more crops in the same field each year. These systems are not mutually exclusive. Two or more may very well be practiced concurrently—cultivated in concentric rings of various lengths of fallow, for example, as in Senegal.

b. R = (number of crop cycles per year × number of years of cultivation × 100) ÷ (number of years of cultivation + number of years of fallow). *Source:* Ruthenberg (1980, 16).

c. These figures are only approximations, the exact numbers depending on location-specific fertility of the soil and agroclimatic conditions. *Sources:* Boserup (1981, 19, 23); Ruthenberg (1980).

d. *Sources:* Ruthenberg (1980); Boserup (1965).

brings a reduction in the period of fallow, and eventually the forest land degenerates to bush savanna. Bush fallow is characterized by cultivation of a plot of land for two to six years followed by six to ten years of fallow. The period of fallow is too short to allow regrowth of the forest. Increasing population density is associated with longer periods of continuous cultivation and shorter fallow periods. Eventually the fallow period becomes too short for the growth of anything but grass. The transition to grass fallow occurs at population density of around sixteen to sixty-four persons per square kilometer. Further increases in population cause a movement to annual cultivation and multiple cropping, the most intensive systems of cultivation.

This movement of the agricultural system from the extensive to the intensive margin is conditioned and sometimes constrained by the climate. The semiarid zones and the tropical highlands are densely populated and more intensively cultivated, but the persistence of forest-fallow cultivation in the subhumid and humid tropics is one example of a climatic constraint. Soils in the humid lowlands generally tend to have poor physical structure and are therefore susceptible to erosion and excessive leaching when cultivated continuously.[2] Forest and bush vegetation protect the soil from the hazard of such degradation. Well-formed soils, such as alluvial and volcanic soils, are exceptions on which permanent production of field crops can be sustained, even with high rainfall, with relatively low levels of degradation.

At the opposite end of the moisture regime, agricultural intensification in the arid zone is constrained by the extremely short growing season. When the intensification of the farming system is constrained by climatic conditions, an outmigration of population to more favorable areas or a lack of immigration into the area can be observed.

Farming Intensity, Cultivation Techniques, and Labor Use

As discussed by Boserup (1965) and Ruthenberg (1980), holding tools constant, the total labor input per hectare on a given crop is correlated positively with the intensity of farming. Table 2-2 presents examples of changes in the use of labor in rice production. The movement from forest fallow to annual cultivation is associated with an increase in total labor input per hectare from 770 hours in Liberia to 3,300 hours in Cameroon. The increase in labor input occurs because of increases in the intensity with which certain tasks, such as land preparation and weeding, must be performed and increases in the number of operations such as manuring and irrigation. Table 2-3 shows how the operations performed increase with the intensification of the farming system.

In the forest-fallow and bush-fallow systems of cultivation, land clearing, planting, and harvesting are the principal tasks. Fire is the most prevalent means of clearing the land. In addition to regenerating the soil through the ashes of burnt vegetation, this way of clearing the land also leaves the soil weed-free.[3] Tree cover is removed, but stumps are allowed to remain in the ground and ensure speedy regeneration of vegetation when the plot is returned to fallow. (In order to use the plow in such a system not only would land have to be cleared, but stumps and

Table 2-2. *Examples of Labor Use with Different Intensities of Rice Cultivation*

	Region and country						
Item	*Gbanga, Liberia*	*Man, Côte d'Ivoire*	*Begora, Ghana*	*Bamunka, Cameroon*	*Ferozepore, India*	*Subang, Java*	*Laguna, Philippines*
Intensity of farming	11	24	40	100	121	200	180
Technique	Hoe	Hoe	Hoe	Hoe	Animal plow	Animal plow	Tractor
			Hours per hectare				
Land clearing	418.4	300.8	665.0	—	—	—	—
Land preparation	—	—	—	714.0	86.4	494.4	73.6
Sowing and planting	107.2	142.4	207.0	536.8	129.6	146.0	80.0
Fertilizing and manuring	—	—	—	—	12.8		
Weeding	36.8	292.0	276.8	113.0		218.0	213.0
Plant protection	44.0	222.0	—	1,393.0	57.6		96.0
Harvesting		218.4		264.0	128.8		
	164.0		280.0			324.4	222.4
Threshing		84.0		280.0	76.8		
Other	—	—	—	—	136.0[a]	70.0	—
Total	770.0	1,259.2	1,432.0	3,300.0	627.2	1,252.0	685.0

— Not applicable.
a. Irrigation.
Source: Ruthenberg (1980).

Table 2-3. *Comparison of Farming Operations in Different Farming Systems*

Operation or situation	*Farming system*				
	Forest fallow	*Bush fallow*	*Short fallow*	*Annual cultivation*	*Multiple cropping*
Land clearing	Fire	Fire	None	None	None
Land preparation and planting	No land preparation; use of digging stick to plant roots and sow seeds	Use of hoe and digging stick to loosen soil	Plow	Animal-drawn plow and tractor	Animal-drawn plow and tractor
Fertilization	Ash; perhaps household refuse for garden plots	Ash; sometimes chitimene techniques;[a] household refuse for garden plots	Animal dung or manure; sometimes composting	Manure; sometimes human waste; composting; cultivation of green manure crops; chemical fertilizers	Manure; sometimes human waste; composting; cultivation of green manure crops; chemical fertilizers
Weeding	Minimal	Required as the length of fallow decreases	Intensive weeding required	Intensive weeding required	Intensive weeding required
Use of animals	None	Animal-drawn plow begins to appear as length of fallow decreases	Plowing, transport, interculture	Plowing, transport, interculture, postharvest tasks, and irrigation	Plowing, transport interculture, postharvest tasks, and irrigation
Seasonality of demand for labor	Minimal	Weeding	Land preparation, weeding, and harvesting	Land preparation, weeding, and harvesting	Acute peak in demand around land preparation, harvest, and postharvest tasks
Supply of fodder	None	Emergence of grazing land	Abundant open grazing	Open grazing restricted to marginal lands and stubble grazing	Intensive fodder management and production of fodder crops

a. To augment the ashes from the bush cover, branches are cut from surrounding trees, carried to the plot of land to be cultivated, and burned to provide extra nutrients for the soil.

roots would have to be removed, a far more arduous task than clearing away the vegetation above the surface.) Clearing land by fire requires very low levels of labor input: 300 to 400 hours per hectare for forest-fallow systems in Liberia and Côte d'Ivoire. Because the ground under tree cover is soft, no further preparation of the land is required before sowing, a task that is easily accomplished with the help of a digging stick or a hand hoe. The forest cover has long suppressed the growth of weeds and few weed seeds remain. Newly cleared land therefore requires almost no weeding or interculture during the first year or two, and the period between planting and harvesting is virtually task-free.

As the fallow period becomes shorter and the land under fallow becomes grassy, fire can no longer be used for preparation of the land. Fire will not get rid of grass roots, so grasses persist through the growing season. The intensive use of a hoe in preparing the land becomes essential. Land preparation and sowing constitute almost 40 percent of the total labor input required for the annual cultivation of rice in Cameroon, and weeding requirements early in the season become increasingly burdensome.

Under forest-fallow and bush-fallow systems, long-term fertility of the soil is maintained by periodic fallowing of land. Renewed vegetative growth on fallow land helps to return fresh organic matter to the topsoil, thus recharging it with supplies of nutrients. Also, burnt ashes return to the soil the nutrients taken up by tree and bush cover. In the zones with lower rainfall, where vegetative cover is less, labor-intensive chitimene techniques may be required: to augment the ashes from the bush cover, branches are cut from surrounding trees, carried to the plot of land to be cultivated, and burned to provide extra nutrients for the plot. Under grass fallow, the supply of nutrients to the soil declines. Accordingly, at this stage the farmer begins to complement fallow periods with additional organic wastes from the household, mainly in the form of vegetative waste and dung from cattle and livestock. As farming intensifies, more labor-intensive fertilizing techniques, such as manuring and eventually composting, evolve.[4]

For the locations visited in Sub-Saharan Africa, figure 2-1 shows the use of organic fertilizer. Under the forest-fallow system no organic fertilizer is used, and under bush fallow there is a move toward the use of kraal dust and one instance of the use of compost, although in the majority of instances no organic fertilizer is used. By the grass-fallow or short-fallow stage, however, there is a marked switch; in eight out of fourteen instances kraal dust or some more intensive matter is used. In most instances of annual or multiple cropping some form of organic

Figure 2-1. *The Use of Organic Fertilizer in the Evolution of Farming Systems*

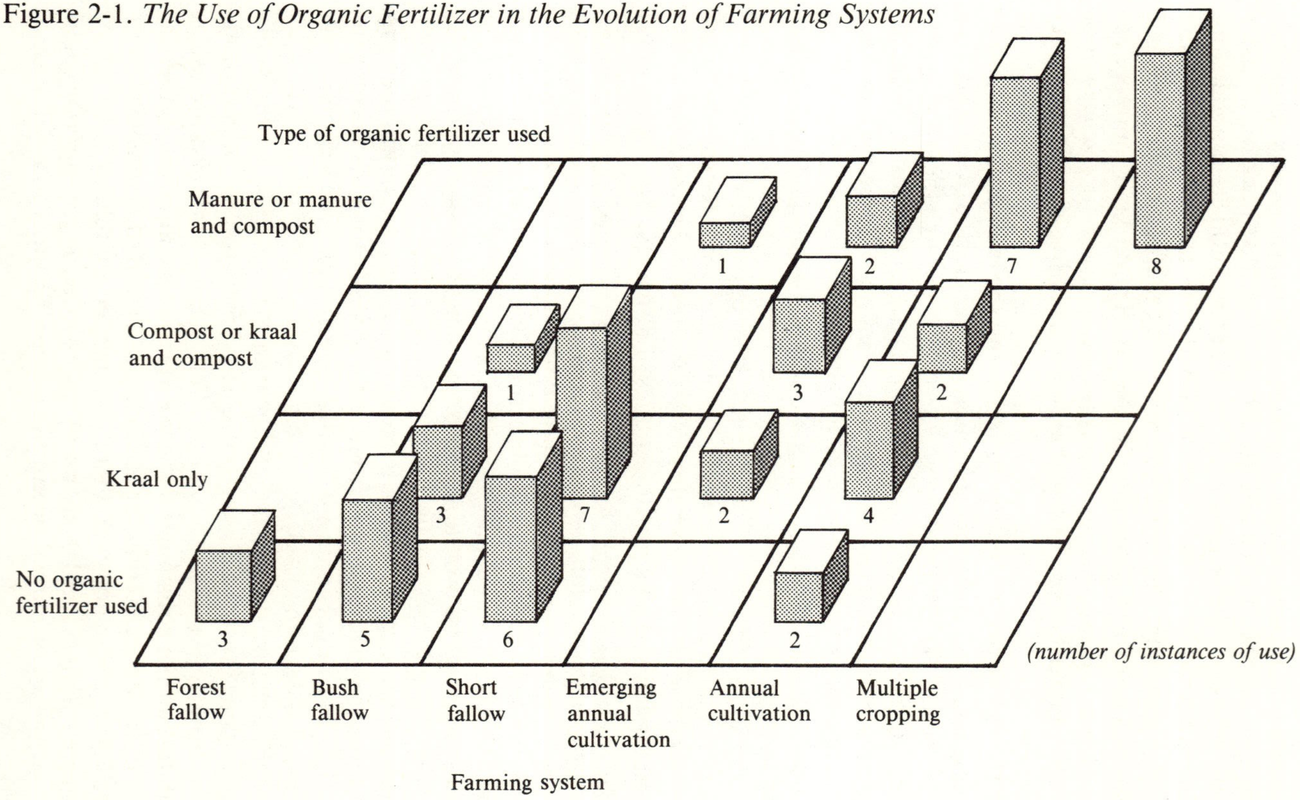

matter, most often animal manure, is used to restore the fertility of the soil. The two exceptions are in the Arusha region of Tanzania, where nothing is yet being done to restore the fertility of the volcanic soils.

Permanent cultivation of land often requires the input of labor for investments in drainage, in leveling or terracing, or in irrigation.[5] Preparation of the land, interculture, and weeding become important labor-intensive tasks. This requirement of labor for crop production can be reduced by a switch to more labor-saving technology, such as the plow for preparation of the land, a cart for transport to the field and to the market, and mechanical mills for reducing the time required for food processing. The transition from the hoe to the plow will be discussed in detail in the next section.

From the Digging Stick to the Plow

The evolution from the hand hoe to the animal-drawn plow is shown in figure 2-2, in which the costs of labor per unit of output using hand-powered and animal-powered systems of cultivation are compared. The progression from one set of tools to the next would take place when the resultant labor-saving benefits exceeded the variable and fixed costs of changing to new tools.

The overhead labor costs in the transition from hand to animal power are the cost of training animals, the cost of destumping and leveling the fields, and the cost of feeding and maintaining the animals year-round. The cost of training the animals is independent of the intensity of farming (line *AB*). The cost of destumping is extremely high when the forest-fallow and early bush-fallow systems are used because of the high density of stumps and the highly developed network of roots. As the length of fallow decreases the costs of destumping decline because the density of trees and roots has been reduced. Destumping requirements disappear at the grass-fallow stage (line *CD*). The costs of feeding and taking care of draft animals—the difference between lines *EF* and *CDB*—is also high during the forest-fallow and early bush-fallow stages because of lack of grazing land and the prevalence of diseases such as trypanosomiasis. As the fallow becomes shorter, grazing land becomes prevalent and trypanosomiasis declines, causing the costs of maintaining draft animals to decline. Beyond the annual cultivation stage, however, grazing land becomes a limiting factor that necessitates the production of fodder crops, which in turn leads to an increase in the cost of feeding and maintaining draft animals. The total cost of using draft animals for land preparation, early-season weeding, and manuring is given by the curve GT_p.

Figure 2-2. *A Comparison of Labor Costs with the Practice of Hand Cultivation and Animal-powered Cultivation*

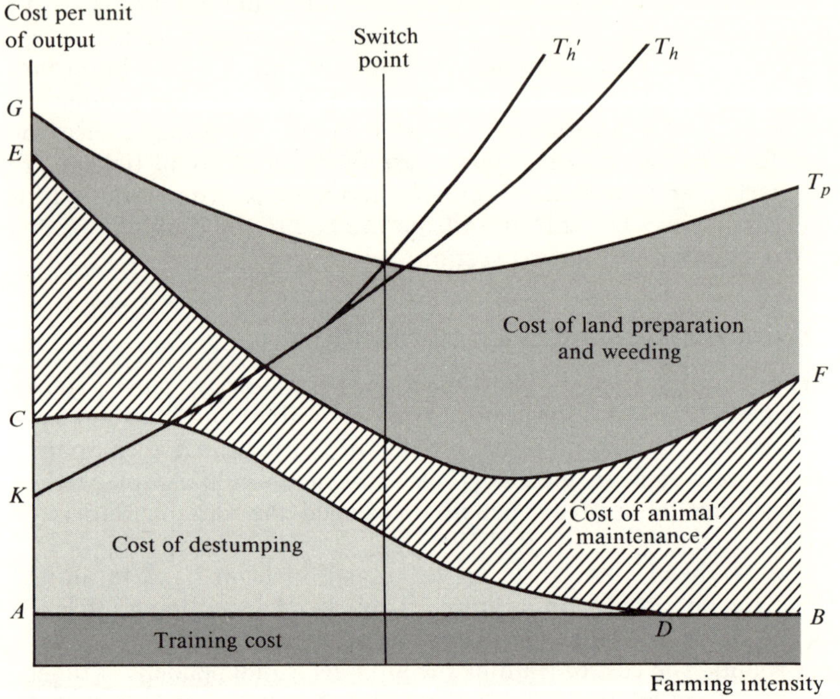

T_p = Total labor costs for land preparation, early season weeding, and manuring, using animal traction

T_h = Labor costs for land preparation and early season weeding, using the hand hoe

$T_h' = T_h$ plus labor costs for maintaining soil fertility without manure from draft animals

Switch point = Farming intensity at which animal traction becomes the dominant technology

As discussed earlier, the cost of labor for cultivation by means of hand tools rises rapidly as farming intensity increases. The effort required for preparation of land, weeding, and maintaining the fertility of the soil increases. T_h shows total labor costs using hand hoes for land preparation and weeding, while T_h' adds in the cost of maintaining the fertility of the soil without the benefit of animal manure.

Animal-drawn plows are the dominant technology at the point at which the costs of hand cultivation exceed the costs of transition to animal power. This switch point is shown in figure 2-2. It is obvious that the figure is a highly stylized representation of the choice between hand hoe and animal draft. In a particular location the choice will be influenced by the type of soil or terrain and the value of animal products. The heavier the soil the greater the advantage of the animal-drawn technology. Heavier soils will shift both the KT_h and the GT_p curves upward but will do so faster for the hand hoe and thereby shift the switch point to the left. The use of animals is typically more difficult on mountain slopes.

In addition to draft, cattle produce meat, milk, and hides. The higher the value of these byproducts, the farther the switch point will be shifted to the left. Up to about seven years of age draft animals add meat of high quality to their weight. When proximity to cities adds good market opportunities for this type of meat, introduction of draft animals can be accelerated.

The benefit to be derived from the use of manure is already incorporated into figure 2-2. Other things being equal, the more difficult it is to produce organic manure without animals, the earlier the shift to draft animals will be made, whereas the availability of cheap fertilizers, by providing a substitute for manure, will tend to favor the persistence of hand-hoe cultivation. To summarize,

- The transition to the plow would not be cost-effective in forest-fallow and bush-fallow systems because the use of the plow would increase weeding requirements and require overhead labor investments for destumping and animal maintenance.
- There is a distinct point in the evolution of agricultural systems at which the use of the plow becomes economical.
- This point will be reached sooner if the soil is difficult to work, if maintaining the fertility of the soil requires a considerable input of labor, or if the market for meat is well developed.

Many of these conditions also prevail when tractor-drawn plows are used. Tractor-drawn plows have not been introduced successfully into Sub-Saharan Africa earlier than the grass-fallow stage. Indeed, as will be seen in chapter 10, it is difficult to bypass the animal-traction stage and move directly to tractors, because the costs of destumping, infrastructure, and learning are much higher in connection with tractor operations than with the animal-drawn plow.

While the use of the plow becomes common by the annual cultivation stage, intensification of the agricultural system does not always lead to the use of the plow. The persistence of the use of the hoe, even under permanent cultivation, can generally be explained by location-specific soil types and terrain. The hill slopes of Sukumaland, for example, have a sandy loam soil that is highly susceptible to erosion. Crop production can be observed on handmade ridges, while at the same time most of the valleys in Sukumaland, which have deep alluvial soils, are cultivated by ox-drawn plows. Highland areas in Kenya, Rwanda, and Burundi are densely populated and intensively cultivated without the use of the plow (Jones and Egli 1984). Under these conditions the GT_p curve in figure 2-2 would lie entirely below the KT_h curve. Some of these exceptions to the general pattern of transition to animal-drawn plows will be discussed in detail in chapter 5.

A frequency chart of the use of tools with the evolution in farming systems, based on the data set from Sub-Saharan Africa, is presented in figure 2-3. In all instances of forest fallow hand tools are used for preparation of the land, while in six of the nine bush-fallow examples hand tools are used. In half the short-fallow instances—seven out of fourteen—animal-draft power is used, in three locations a combination of animal and tractor power is used, while in only four the hand hoe continues to be relied on exclusively. The dominance of mechanical tillage—by means of animal draft or tractors—becomes more prominent as short fallow is replaced by permanent cultivation. Of the thirty examples in which permanent cultivation is emerging or established, twenty-seven reported using animal draft or a combination of animal draft and tractors for tillage. The exceptions, where hand tillage persists, are the hill slopes of Sukumaland, agricultural areas surrounding Kano city, and the Yatenga region of Burkina Faso. In the last two instances the light sandy soils are easy to work by hand and the short rainy season presents farmers with a severe problem of capacity utilization, an issue that will be discussed in detail in chapter 4.

Areas in Which Destumping Is Not a Major Constraint

While it is generally true that the transition to the plow would not be cost-effective in forest-fallow and bush-fallow systems because large

Figure 2-3. *The Use of Tools in the Evolution of Farming Systems*

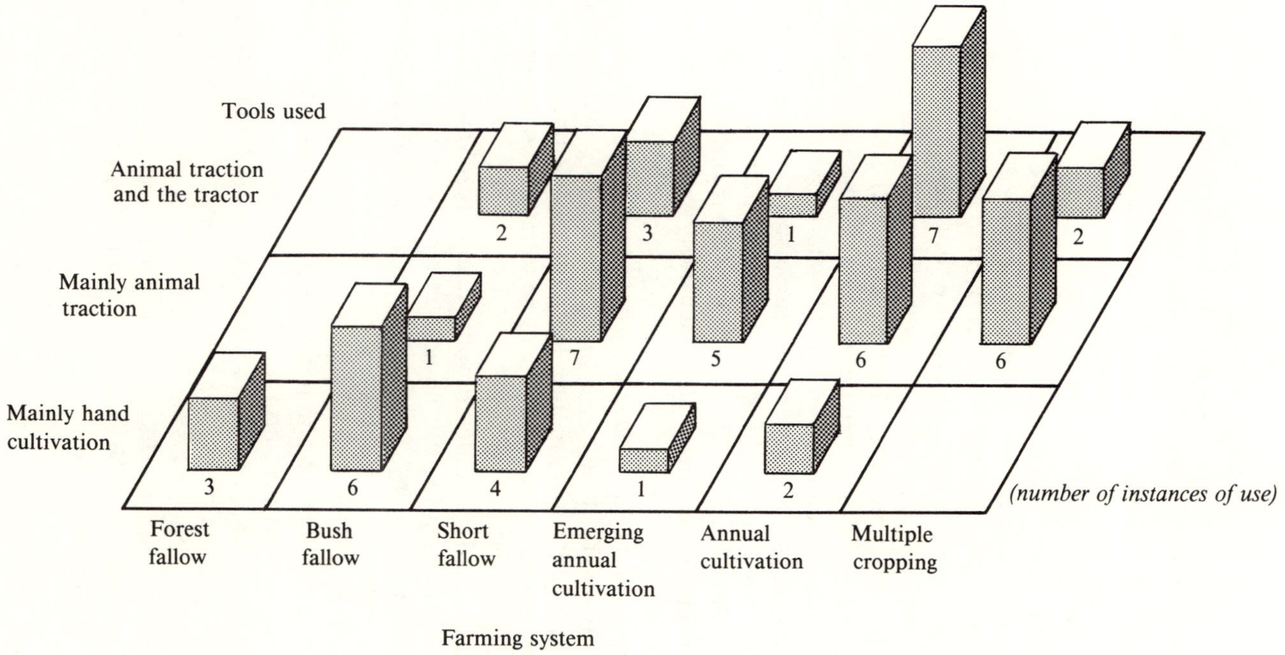

overhead investments in labor for destumping of fields would be required, there are areas, such as the flood plains and grassy savannas that are naturally free of stumps, where the transition to the plow can be made more rapidly.

Flood Plains and Riverbeds

Flood plains and riverbeds are generally characterized by deep, heavy soils that tend to become waterlogged or flooded periodically. These soils are often extremely difficult to work by hand, and where the density of population is low they are either left for dry season grazing or are completely unused. Once a mechanical source of power and the plow have become available, cultivation of these lands can expand rapidly. Rice is the preferred crop because of the abundant availability of water and the tolerance of rice to waterlogging or flooding.

The valley bottomlands in Sukumaland have heavy black clay soils, the "black mbugas," while the soils of the hill slopes are very light. Until the 1940s the valley bottomlands were left mainly for cattle grazing and the hill slopes were cultivated intensively by hand. Rounce (1949) summarizes the situation as follows: "The use of the plow should be strongly encouraged on the 'mbuga' soil, only a very small proportion of which is cultivated by hand owing to the short period during which it can be tilled and to its heavy nature. The more the bottom lands are cultivated by the plow the more there will be to fallow on the light soils on the slopes" (p. 15).

Rounce estimated that it would take sixty-seven workdays for preparation and planting of one acre of rice by hand on the mbuga soil, twenty-two workdays for groundnuts, twenty workdays for cotton, and eighteen workdays for maize grown on handmade ridges on the slopes (p. 6, table 6). The colonial authorities accordingly encouraged plowing in the valley bottom, and today intensive production of rice, using ox-drawn plows, is the norm on all these lands.

Similar examples of the rapid spread of mechanical tillage for flooded rice production can be found in Spencer (1981) for the riverain grasslands of Sierra Leone, Tuluy (1981) for the Upper Casamance in Senegal, McIntire (1981) for the Delta areas of Mali, and Kjoerby (1983) for the Lake Nyasa flood plain in Tanzania and the Odienné region of Côte d'Ivoire.

Grassy Savannas of Semiarid Africa

Several parts of semiarid Africa are covered with perennial grasses interspersed with bush and small trees. Until recently, these grassy savannas had been pastoral areas with almost no land under cultivation. The main constraint to cultivation of these lands by hand is the large amount of labor required for removing the grass cover before planting. Fire, as mentioned earlier, does not remove the grass roots, which can be more easily controlled when the soil is turned over by the plow.

The introduction of the plow often enables a rapid increase in the amount of grassy savanna land under cultivation. The diffusion of animal-drawn plows is made easier by the availability of large herds of cattle, animal husbandry skills, and the few stumps in the fields.

The Kyoga basin in Uganda is a good example of the spontaneous spread of animal traction in the grassy savannas. Before the 1900s this area was inhabited mainly by nomadic and seminomadic pastoralists who owned large herds of cattle and subsisted largely on milk, blood, and meat. The plow was introduced by the colonial government in 1910, and by 1936–37 there were 15,388 plows in Teso district alone (Tothill 1940, 55–56). By 1969 there were 70,000, and virtually all the new land was being opened by the plow in 1980 (Carr 1982).

Trypanosomiasis as a Constraint to Agricultural Intensification

African animal trypanosomiasis is a chronic disease that is usually fatal to domestic animals. It is transmitted by the tsetse fly (*Glossina* sp.), which lives exclusively in tropical Africa. There are three species of Glossina: *palpalis, morsitans,* and *fusca.* The palpalis lives in forest galleries or in the marginal areas of forests and the fusca is generally associated with equatorial forests, while most of the morsitans live in wooded savanna (FAO 1983b).

Map 2-1 shows the extent of tsetse infestation in Sub-Saharan Africa. The presence of the tsetse fly is felt throughout virtually all the humid and subhumid zones and extends in riparian vegetation even into the arid zone and the fringes of the highlands (Pratt 1984, 32). Tsetse is most common above the 800-millimeter isohyet, though in riparian vegetation and wet depressions it can exist where annual rainfall is less than 600 millimeters or even 400 millimeters. Tsetse flies generally disappear at altitudes higher than 1,500 meters (LeHouerou and Popov 1981, 7).

Trypanosomiasis is a serious threat to livestock holdings in the humid and subhumid zones of Africa. Jahnke (1982) estimates that there are fewer than three head of cattle per square kilometer in the humid tropics, while there are eleven per square kilometer in the

Map 2-1. *Ecological Zones of Tropical Africa and Area of Tsetse Infestation*

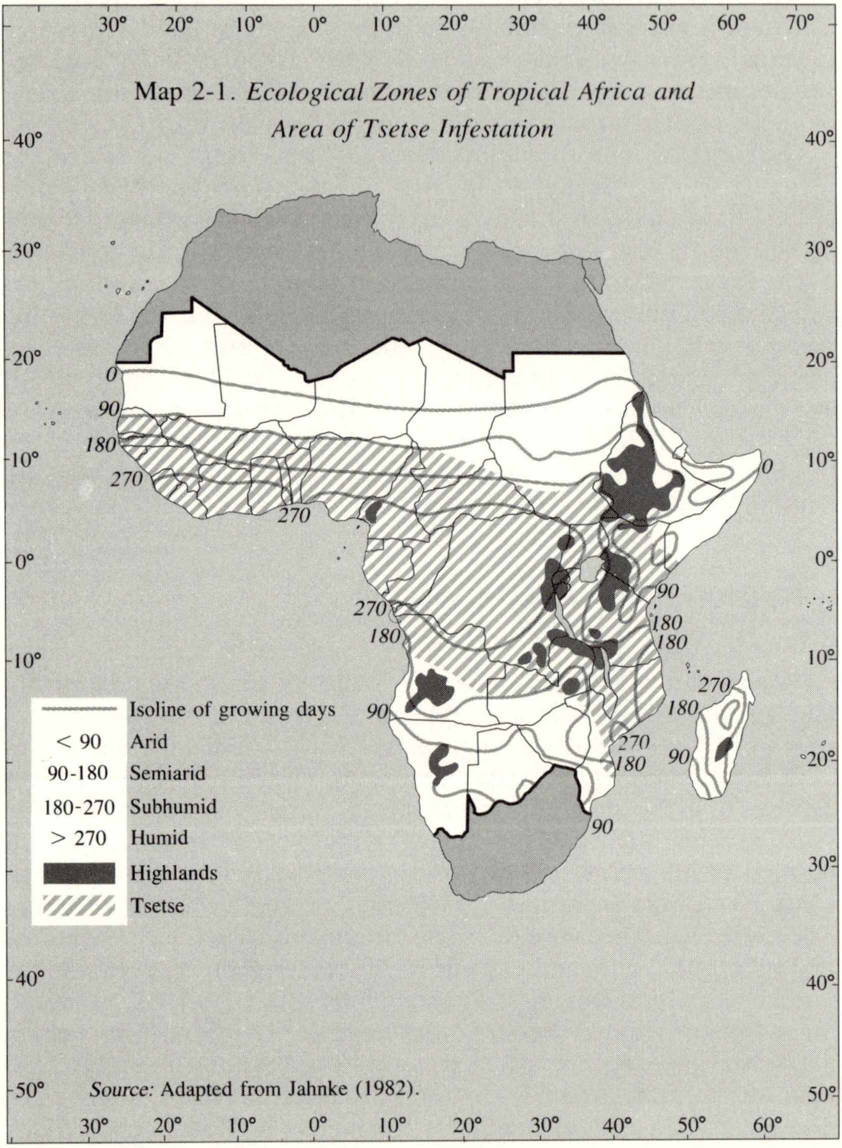

Source: Adapted from Jahnke (1982).

semiarid tropics and twenty-nine per square kilometer in the highlands of Africa. Zebu cattle are highly susceptible to trypanosomiasis, while the small humpless cattle of West Africa, such as the N'dama and the Baoule, tolerate the disease well.

Population Density and the Habitat of the Tsetse Fly

The tsetse fly is shade loving and can survive only under forest or bush cover. A recession in this vegetative cover leads to a recession in infestation by the tsetse fly and therefore to a change in the incidence of disease. Increases in population density lead to a reduction in forest cover as larger areas are brought under cultivation, and this acts as a natural control mechanism against the fly. This correlation between population density, intensity of land use, and abatement of the tsetse fly has been observed in several parts of Sub-Saharan Africa (Jahnke 1976, Ford 1971, Tiffen 1976, Bourn 1983, and Jerve 1982, among others).

Tiffen discusses two contrasting cases in Nigeria. In the Majnyoru valley intensive programs for eradication of the tsetse fly were carried out in 1926–30 and 1962. Soil surveys subsequently showed the area to be of low agricultural potential, and for that reason agricultural populations remained low. Once the government-sponsored slashing had stopped, the tsetse fly returned. In contrast, Yamaltu district, with good soils and good marketing opportunities, attracted immigrants in the early 1950s. The consequent extension of the cultivated area caused the tsetse fly to retreat, without any government-sponsored eradication program.

Low population density rather than tsetse fly infestation can be considered the reason for low intensity of land use. Given suitable soil conditions, increases in population density have led to the expansion of agricultural land into tsetse-infested areas. Where the condition of the soil is marginal, however, population growth leads to emigration rather than to expansion of cultivated area. If the tsetse fly were the most important constraint and if tillage were profitable, use of the tractor would be expected to be prevalent in these areas, but it is not. Nevertheless, at the forest margin the tsetse fly is an important constraint to the use of animal traction. Where population density increases farming intensity at the forest margin, N'dama and other trypano-tolerant breeds of cattle could be used to provide traction power.

Sierra Leone is a good illustration of the use of N'dama cattle; much of the agricultural land in the country lies between the forests and the savannas. Starkey (1982) reports that N'dama cattle were successfully introduced for draft purposes at the forest margin. They were used—and continue to be used—primarily for cultivation of rice in depressions and swamps, where, as discussed earlier, tree stumps are no constraint on the use of the plow. But use of the plow is spreading only slowly into the upland areas, where stumps continue to be a constraint.

Examples of unsuccessful animal-traction projects using N'dama cattle can also be found. Compagnie Ivoirienne pour le Développement des Textiles (CIDT) has made several attempts to introduce N'dama cattle for draft in Morondo, which is located in the sparsely settled forest zone of Côte d'Ivoire. These attempts have failed because the farm population still practices shifting cultivation on uplands, a farming system in which the plow would be inappropriate even if trypanosomiasis were absent.

Implications for Tsetse Control

Jahnke (1976) summarizes the techniques available for controlling trypanosomiasis. The first and best measure is intensification of land use. If intensification is not possible, no other control technique will eliminate the threat permanently (Jahnke 1976, 112).

Techniques for control of the tsetse fly are most effective where an expanding population begins to encroach on a forest area that is infested by the fly. Preventive measures such as selective clearing of bush, application of insecticides, and vaccination of cattle are therefore most useful at the forest margin.

Notes

1. The precise definition is: R = (number of crop cycles a year × number of years of cultivation × 100) ÷ (number of years of cultivation + number of years of fallow)—that is, the frequency with which a plot of land is used. R may vary from 0 to 2 or 3 in systems in which multiple cropping is practiced.

2. Whenever remunerative markets are available, shifting cultivators in the humid and subhumid tropics, where rainfall is abundant, tend to plant tree crops such as rubber, banana, coconut, and cocoa, since these, once established, require less weeding and protect the soils from erosion or excessive leaching (Ruthenberg 1980, 45).

3. Miracle (1967) identifies eleven different methods of clearing land with fire. Clearing the land is not required when short-fallow and permanent systems of cultivation are practiced, since the fields are no longer under tree or bush cover.

4. The inhabitants of densely populated Ukara Island in Lake Victoria, Tanzania, laboriously collected three tons of manure a year from each adult head of cattle and transported it to the fields by headloads (Allan 1965, 201). In the villages near Kano city, Nigeria, farm manure was supplemented with night soil transported from the city (Gleave and White 1969, 284). In areas in which livestock herding has traditionally been separated from crop farming contracts tend to be made between farmers and herders for the exchange of cattle droppings for stubble grazing (Toulin 1983).

5. The densely populated hilltop refuges of Africa provide several historic examples of ridging and tie-ridging, silt traps, and elaborate systems of stone-walled terraces. Such investments were made before the colonial period in Jos plateau, Nigeria; the Mandara Mountains, Cameroon; the Kikuyu highlands, Kenya; and Rwanda.

Chapter 3

The Driving Force of Intensification

Agricultural intensification is caused by a decrease in the cultivable area per capita when the population density increases because of the growth or concentration of population in particular regions (see table 3-1). The causes of concentration are both historic—tribal war, slave trade, and so on—and contemporary—land-use policies, lower costs of transport, and so on. Concentration of population also occurs in areas endowed with high altitude, fertile soil, and other favorable conditions.

Though the causes of increases in population density are varied, the consequences—reduction in cultivable land per capita and the intensification of the agricultural system—are fairly similar in most locations. In addition to reducing the length of fallow on land already in use, whether cultivated or fallow, growing population density also leads to the colonization of lands—such as swamplands—formerly unused for agricultural purposes. The expansion to marginal lands will be discussed in detail in chapter 4.

Natural Population Growth and Disease Control in Recent Times

Ukara Island in Lake Victoria provides a celebrated example of the process of intensification that follows a natural increase in population. The Wakara moved to a small island in Lake Victoria at some time around the fifteenth century to avoid their warlike neighbors in northern Tanzania. The island is supposed to have been densely settled by the seventeenth century, so its inhabitants have had a long history of land shortage and population pressure. As early as the nineteenth century the agricultural practices there included manuring, production of fodder crops, erosion control measures such as terracing, and irrigated

Table 3-1. *Causes and Consequences of Increased Population Density*

Condition	*Cause*	*Consequence*	*Action taken*
Natural growth of population	Improved public health and lack of emigration		*Reduction in fallow periods* Movement from shifting to permanent cultivation
Soil fertility	Immigration to capture the benefits of higher returns to labor input		*Mechanization* Plowing—where agroclimatic and soil conditions make it profitable
Transport facilities[a]	Immigration to capture the benefits of reduced transport costs		Transport—where markets exist for food and other crops
Urban demand[a]	Immigration to capture the benefits of proximity to markets	Reduction in available area per capita	Milling—in response to higher opportunity cost of time for female household members
Health	Immigration to cooler highlands to avoid malaria and the tsetse fly		*Investments in land* For soil fertility, drainage, terracing; increase in the marginal lands brought under cultivation
Historic	Immigration to inaccessible highlands to avoid tribal warfare and the slave trade		
Land laws, rights	Restrictions on the right to open new land		*Land rights* From communal to private ownership of land

a. With improved transport facilities and urban demand and in the absence of immigration, the area under cultivation will expand.

rice farming of the lowlands and lakeshore regions. The island was also characterized by individual ownership of all land (Ludwig 1968).

After the colonial powers had succeeded in controlling East Africa, many of the Wakara migrated to the regions bordering the southeastern part of Lake Victoria. There they abandoned the advanced methods of crop husbandry and resumed shifting cultivation.

In Rwanda and Burundi improvements in public health since 1900 have led to a sharp decline in death rates and a consequent increase in population density. The present density on the High Plateau is around 250 persons a square kilometer (Jones and Egli 1984). The consequent intensification of agricultural production has led to a shortening of grazed fallows by the planting of a higher proportion of land each year; the cultivation of steeper slopes; and drainage of the swamps so that they can be used for dry-season crops.

Several other examples of population growth brought about by improved public health services and the consequent changes in farming systems can be found in both East and West Africa; see, for example, Almy (1979).

Immigration into Areas with Fertile Soils

For a given density of population the marginal productivity of labor is higher on more fertile soils, so immigration from areas less well endowed and consequent reductions in cultivable areas per capita would be expected. Ada district, Ethiopia (Makkonen and Assamenew 1978), Nyanza province, Kenya (Humphrey 1947), and the southern province of Zambia (Trapnell and Clothier 1937) are a few examples of fertile areas that are densely populated and intensively cultivated. Maragoli and Marama are neighboring divisions in the southern part of Nyanza province. The land in Maragoli is extremely fertile, that in Marama less fertile. Immigration into Nyanza province from the southern part of Lake Victoria took place around the eighteenth century, and by the 1940s Maragoli had a population density of 306 per square kilometer. The land was then cultivated intensively with two main crops a year. In Marama, population density was less than 94 per square kilometer, only one main crop was grown a year, and there were still some fallow land and ample communal grazing. A comparison of farming systems of the two areas is given in table 3-2. Today, the two areas would be expected to be at an equal level of intensity because of natural growth of population in Marama and immigration from Maragoli and other more densely populated areas.

Table 3-2. *The Relation between Fertility of the Soil and Agricultural Intensity, Nyanza Province, Kenya, during the 1940s*

Variable	*Maragoli*	*Marama*
Fertility of the soil	Very high	Marginal
Density of population	306 persons per square kilometer	93.5 persons per square kilometer
Cultivable area per family	1.5 hectares	5.4 hectares
Land rights	Individual ownership prevalent, field boundaries common	Communal ownership, individual cultivation rights
Grazing rights	Private grazing land and restrictions on stubble grazing	Communal grazing land
Cropping pattern	Two main crops a year, no fallow land	One main crop a year, short fallow
Use of manure	Very common	Emerging

Source: Humphrey (1947).

Immigration into High Altitudes

High altitudes have long been preferred for human settlement because of their better climate and lower incidence of disease, notably malaria and sleeping sickness. Concentrations of population in the Ethiopian and Kenyan highlands and the Kigezi district of Uganda are noteworthy examples of this phenomenon. The majority of the population of the Kigezi district is concentrated on the Birunga mountain range at altitudes of 1,800 to 2,250 meters, while the malaria-infested lowlands are sparsely populated. Population growth is greater in the highlands because of lower levels of disease and therefore sharply lower death rates. In 1944, the density of population was about 144 per square kilometer and today it is about 280. As population density increased cultivation proceeded, first upward, toward the summits of hills, then toward the valley. By 1946 almost all the summits were being cultivated and cultivation of the bottomland had begun (Purseglove 1946). Similar intensification was observed in the high altitudes of North Kavirondo district, Kenya (Wagner 1949, 9) and the Kikuyu highlands of Kenya (Humphrey 1945).

Improved Transport Facilities

Given suitable soil conditions, areas with better access to markets will be more intensively cultivated. Reduced costs of transport give access to urban, interregional, and international markets. Intensification occurs for two reasons: (1) Higher prices and elastic demand for exportables implies that marginal utility of effort increases; farmers in the region will therefore begin cultivating larger areas. (2) Higher returns on labor encourage immigration into the area from neighboring regions where costs of transport are higher.

During the 1830s merchants in the Senegambian trading posts discovered that African groundnuts could be sold in Europe and the United States. Production for export began to spread rapidly during the 1840s along the Gambia and Saalum rivers, where efficient river transport had existed for centuries. The lower Gambia valley was sparsely populated, and by the early 1840s farm families had migrated from 300 to 400 miles away to take up land near the Gambia. The immigrant farmers hired land from the heads of local families, sometimes paying the rent with their labor. By 1848 most of the peanuts exported from the Gambia were produced by the immigrant farmers (Curtin and others 1978, 507).

The story of peanut production in Senegal is quite similar. During the 1870s the government of Senegal built a railroad from Dakar to Saint Louis at the mouth of the Senegal River, passing through some of the best peanut-growing territory, which was until then sparsely cultivated. With the advent of the railway peanut production boomed. By 1900 exports through French-controlled ports had reached 140,000 metric tons (Curtin and others 1978, 508). Baier (1980) also cites the case of the Kano region in Nigeria, where the coming of the railways brought an expansion of the area cultivated in groundnut.

There are several other examples of intensification induced by improved transport facilities in East Africa. Intensive production of maize in fertile parts of Kenya and Zambia was a direct consequence of the railways built by colonial governments. The railway from Mombasa reached Kisumu on Lake Victoria in 1901 and stimulated a rapid increase in the production of cotton in Uganda. Until that time headloading of cotton to the coast had cost an estimated £200 a ton; on the railroad it cost £2.40 a ton. As a consequence cotton became Uganda's chief export, rising from £43,000 in 1903–4 to £307,000 in 1910–11, more than half the total exports. By the early years of World War I cotton had reached 70 percent of total exports. Intensive production of cotton drew migrant labor into Uganda; by 1946 the estimated flow was 140,000 from Rwanda and Burundi alone. Many of the migrants settled permanently in the province (Curtin and others 1978, 510).

By the late 1920s motor transport had had an equally impressive effect on agricultural production in East Africa. Motor transport was introduced to Sukumaland during the 1920s, and by 1926 head porterage had almost disappeared from Mwanza district. The area planted in cotton spread rapidly, especially along the newly built roads. Between 1924 and 1940 some 10,000 square kilometers of Sukumaland were cleared of tsetse fly and opened up for farming, out of which only 2,000 square kilometers were officially cleared by the government, the rest by farmers migrating into the area. By the late 1930s plows had been widely adopted in Sukumaland as the area planted in cotton, maize, and rice had expanded in response to better transport facilities and greater demands for food by local gold miners (Iliffe 1979, 287–88).

It is important to remember that intensive farming induced by reduced transport costs also depends on suitable soil conditions. Where soils are less fertile or difficult to work, better transport facilities do not cause an expansion of the area under cultivation. The railway line in Zambia, for example, runs from the copper fields in the north to the southern tip of the country, but intensified production of maize can be observed only along the railway in the southern province and the central province south of Lusaka. Commercial production of maize along the railway in Kenya is similarly concentrated in the areas in which potential production is high. The intensification of farming in areas such as the Machakos and Embu districts, where the potential is only moderate, is a more recent phenomenon brought about by population growth and the migration of population from the densely settled high-potential areas in search of land. Agricultural intensification in response to improved access to markets could take place even where the density of population is low, with individual farmers expanding the area given to marketed crops. The consequences of intensification in these circumstances do not differ sharply from those in areas where the density of population is high.

The Effects of Urban Centers

Agricultural land surrounding urban centers tends to be intensively cultivated, mainly in vegetables and other food crops. Such intensification is in response to urban demand and low transport costs. Production of vegetables around Nairobi, Lagos, and Abidjan and cereal production in Ada district near Addis Ababa are examples of this response to urban demand.

Agricultural production outside Kano city in Nigeria is a good example. For more than 300 years Kano, formerly an end point of the trans-Saharan caravan trade, has been the commercial center of Hausaland. Its importance has increased since the construction of the railways from Lagos and Port Harcourt. Its population has grown from

35,000 in the middle of the nineteenth century to 50,000 in 1921, 130,000 in 1952, and well over 250,000 in the early 1980s. In its vicinity lies the most heavily settled region in the Sudanian zone. Within fifty kilometers of the city population density exceeds 250 persons per square kilometer; practically no land lies fallow and grazing commons have virtually disappeared. There is less than half a hectare of cultivable land per capita.

Millet, groundnuts, guinea corn, and cowpeas are grown on the uplands, while in the depressions and along the water courses small irrigated plots of onions, garlic, wheat, rice, and vegetables are grown. Intensive manuring techniques are common: goats, sheep, and donkeys are kept tethered and their droppings used for manure. Other sources of manure are the composted wastes from the Kano slaughterhouse, waste from old pit latrines, and mud from the beds of streams that pass through the city and act as sewers (Grove 1961, 131). Composting of waste from townships and slaughterhouses in Kampala and Nairobi during the colonial period has also been documented (Hailey 1957).

Tribal War or Slave Trade

In Sub-Saharan Africa, in an effort to avoid intertribal strife or slave trade, population historically became concentrated in highlands to which access was difficult. In East Africa such concentrations occurred in response to the Masai domination in the nineteenth century. The Masai sphere of influence extended from 2 degrees north to 6 degrees south and from 34 to 40 degrees east (Southall 1961, 61). As Masai power expanded, several tribes took to the mountains of the Great Rift valley to avoid confrontation with them. The case of the Iraqw is typical of the subsequent changes in farming systems.

By the late nineteenth century the Iraqw people had occupied a concentrated settlement at Kainam on the western rim of the Great Rift valley. Prevented from expanding by the Masai, they developed an intensive agriculture, with manuring, soil conservation, private ownership of land, and extremely unequal possession of cattle. During the early 1890s the Iraqw expanded, pushing northward and settling below the Rift wall (Iliffe, 1979, 351). In these areas herds increased quickly and the sophisticated agriculture of Kainam was quickly abandoned. Other places in which population became concentrated in response to Masai expansion are Arusha province, Tanzania; Meru, Kenya; Chagga, Kenya; the Pare mountains, Tanzania; Sambaa, Kenya; Kikuyu, Kenya; Kamba, Nigeria; and Teita, Kenya (Southall 1961, 61).

The concentrations of population and the growth of well-developed political organizations and intensive agricultural systems on the high plateaus of Rwanda and Burundi were in response to the incursions of Arab slave traders into the area (Jones and Egli 1984). Gamo society took to the southern highlands of Ethiopia to avoid intertribal war in which the vanquished were taken as slaves (Abeles 1981, 38–39).

The common feature in all these cases was the intensive cultivation of small plots of land with extremely high inputs of labor for investments such as terracing and for maintaining soil fertility. Wherever these societies were allowed to spread out of their concentrated settlements, they invariably abandoned the intensive practices and reverted to shifting cultivation.

Land Laws and Land Rights

Policies that restrict the acquisition of new agricultural land lead to an intensification of the agricultural system. Both the right to use traditional fallow land and the right to colonize new land as population expands are restricted; as a consequence, the amount of land available per capita and the length of fallow periods are reduced. In East Africa such intensification was caused by colonial land policies whereby large tracts of land were alienated to European farmers.

Under German rule in Tanzania, the lands on the lower slopes of Mt. Kilimanjaro and Mt. Meru were alienated to European settlers. The colonial policy did not dispossess the indigenous population of land already under cultivation but restricted their right to open virgin land and land that had long lain fallow. Until that time the land on the lower slopes had been used for cereal production, with fairly long fallow periods, while the fallow periods on the upper slopes were short. The ring of white settlers on the lower slopes increased the pressure on the land held by the indigenous population, causing a switch to permanent cultivation of all the land on the upper slopes. Reduction in arable land per capita also caused many in the indigenous population to migrate to the settler estates in search of employment (Hyden 1980).

The encouragement of large-scale production of maize by European farmers in the Southern province of Zambia caused similar intensification of farming on indigenous farms. In the Mazabuka district, of a total of 5,587 square kilometers of cultivable land, 3,129 square kilometers were declared European land and the indigenous population was concentrated on 2,458 square kilometers. Intensification came about because of a loss of traditional fallow land and the migration of workers into the district from neighboring districts off the railway line to seek employment on European farms.

Evidence of Agricultural Intensification

The foregoing discussion leads to the broad generalization that for given climatic conditions, increases in density of population and improvements in access to markets gradually cause a transformation of the agricultural system from forest fallow to annual cultivation and eventually to multiple cropping. Empirical support for this generalization is provided by two-way frequency tables using our data set from Sub-Saharan Africa.

Figure 3-1 presents the positive relation between population density and farming intensity. Forest-fallow and bush-fallow systems are predominant where population density is sparse—where there are fewer

Figure 3-1. *Population Density and Agricultural Intensification*

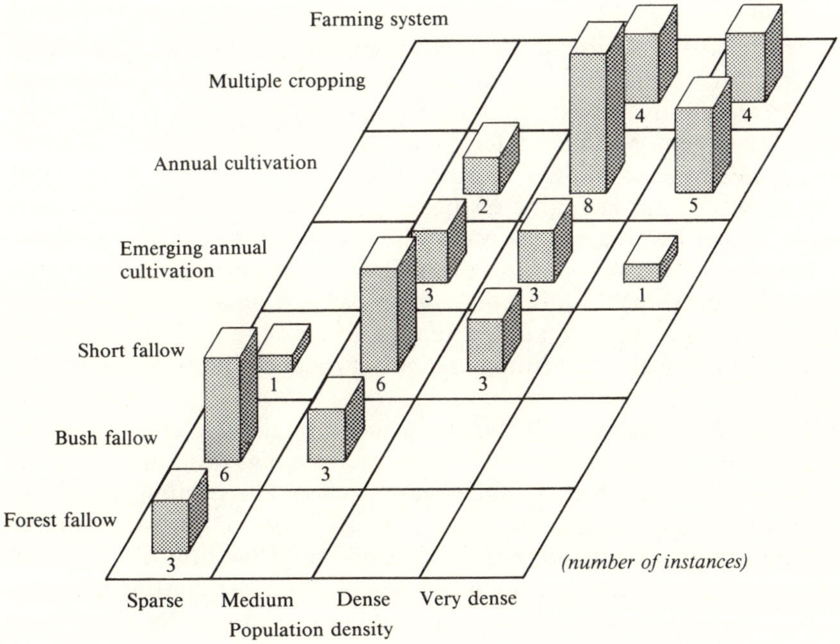

than fifteen persons per square kilometer. These systems of farming are practiced in nine of the ten sparsely populated areas in our sample; the single exception is under short fallow. In the medium-density group—sixteen to twenty persons per square kilometer—the transition to more intensive systems of farming begins to be seen. Here nine out of fourteen instances fall in the short-fallow and emerging annual cultivation categories; there is no forest-fallow cultivation at this density level, and bush-fallow is practiced in only three instances. In the densely populated group—fifty to a hundred persons per square kilometer—cultivation of annual and multiple crops is well established in twelve of eighteen instances in these systems and there are three instances of emerging annual cultivation. Three remaining instances of short-fallow cultivation are found, presumably at the lower end of the densely populated group. In nine of the ten instances in the very dense population group—more than 100 persons per square kilometer—annual or multiple-crop cultivation is practiced, the single exception being one of emerging annual cultivation.

The classification of our examples shows a definite trend toward agricultural intensification as population density increases. During neither our field visits nor our survey of the literature have we come across any examples of sparsely populated areas under annual cultivation or very densely populated areas under forest-fallow or bush-fallow systems.

The positive relation between the ease of access to markets and farming intensity is shown in figure 3-2. For a given population density an improvement in access to markets causes further intensification of the farming system. With poor access to markets extensive forms of farming such as forest-fallow and bush-fallow are usually practiced. With fair access to markets nine of the sixteen instances were observed to have intensities greater than bush fallow. Among those whose access to markets is good, twenty-two of the twenty-four examples cultivate at intensities of short fallow and above. Where access to markets is excellent, nine of the ten examples are under annual or multiple-crop cultivation and the remaining case is approaching permanent cultivation.

Figure 3-3 illustrates the way in which climatic variables constrain the process of intensification. With annual average rainfall of less than 750 millimeters, which includes zones where the rainfall is low, the semiarid zones, and the arid zones, forest-fallow and bush-fallow cultivation are not practiced. This is because the slow rate of vegetative regrowth under arid conditions does not permit regeneration of the forest even at low population density, and cultivation is often feasible on

Figure 3-2. *Market Access and Agricultural Intensification*

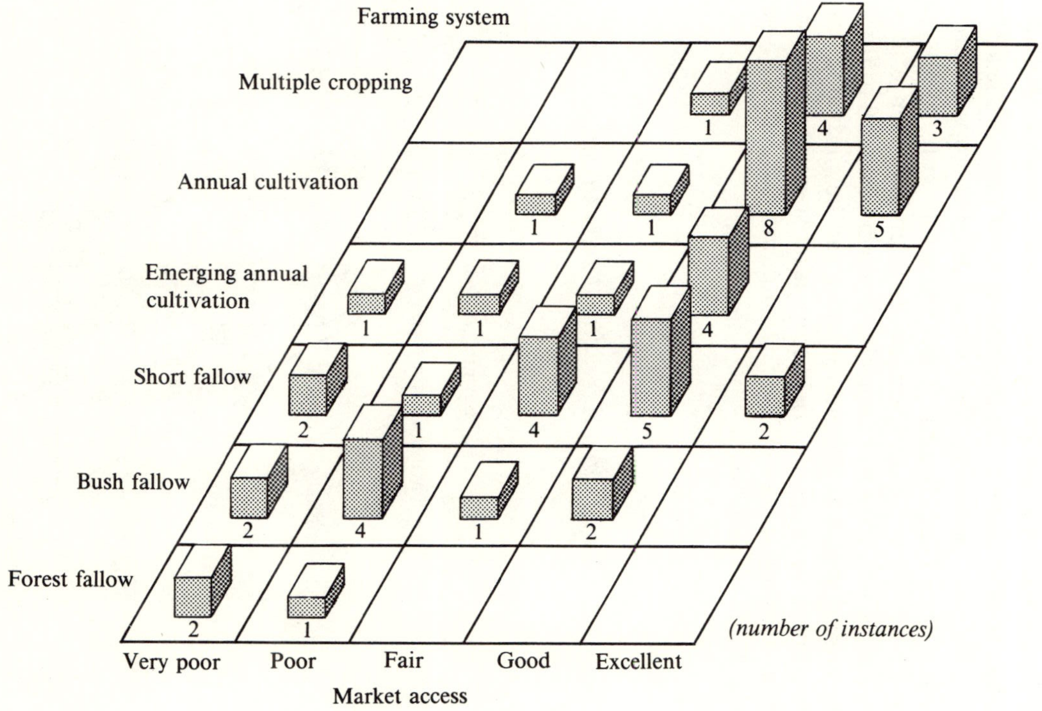

only a small part of the land, namely the lower slopes and depressions, where the capacity of the soil to hold moisture is sufficient to sustain a crop. Where rainfall is high—more than 1,200 millimeters annually—as in the humid tropics, a predominance of forest-fallow and bush-fallow cultivation is to be seen. Because of intensive leaching and soil acidification, permanent cultivation of field crops is hard to sustain under humid conditions. The intensively cultivated humid areas shown in figure 3-3 are from the highlands of Kenya and Ethiopia, which do not suffer from leaching and acidification to the extent that the humid lowlands do and therefore can be cultivated permanently. Permanent cultivation of field crops is most easily sustained in the medium rainfall zones—751–1,200 millimeters annually—and this is consistent with our data.

Figure 3-3. *Climate and Agricultural Intensification*

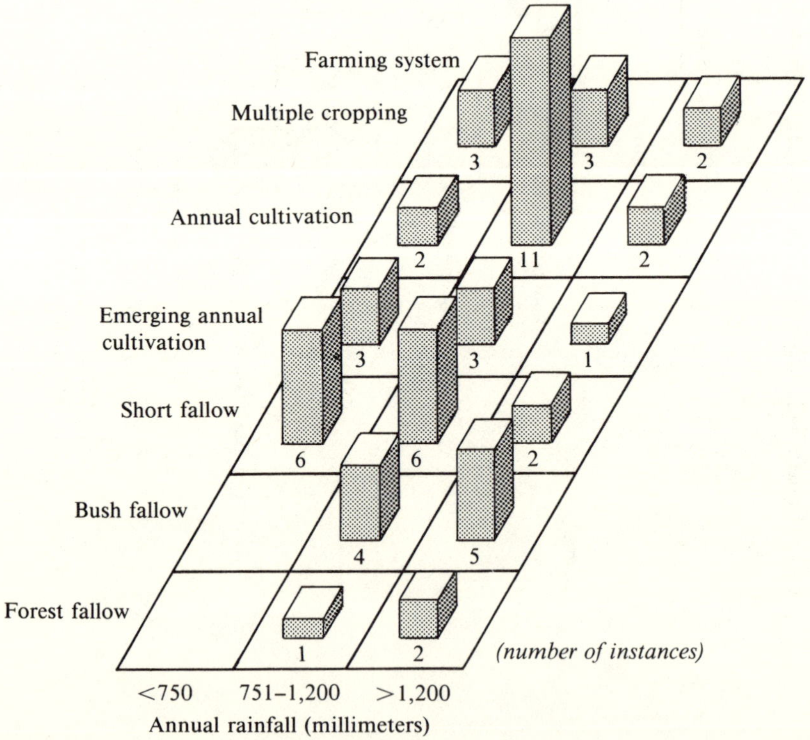

Agricultural Intensification and Changes in Sources of Power

As discussed in chapter 2, use of the plow is closely associated with the intensity of the agricultural system and emerges at the late bush-fallow and early grass-fallow stages, generally not before. The Mumbwa district of Zambia, for example, which is off the railway line, was until recently under forest-fallow cultivation. When the search for land accelerated migration from districts on the rail line, fallow periods were reduced and ownership of cattle and cultivation with oxen emerged.

Table 3-3. *Permanent Systems of Cultivation and Operations Mechanized*

Location	*Reason for concentration or intensification of the population*	*Operations mechanized*	
		Land preparation	*Transport*
Ethiopian highlands	Health and security	x	x
Kenyan highlands	Health and security	—	x
Rwanda and Burundi	Health and security	—	x
Mt. Kilimanjaro, Tanzania	Colonial land policy	—	x
Sukumaland, Tanzania	Transport	x	x
Ukara Island, Tanzania	Population growth	—	—
Arusha province, Tanzania	Soil fertility	x	x
Kigezi district, Uganda	Health	—	x
Mazabuka district, Zambia	Colonial land policy and transport	x	x
Mumbwa district, Zambia	Recent migration	x	x
Kano region, Nigeria	Transport	x	x
Kano city, Nigeria	Urban demand	—	x
Senegal, groundnut basin	Transport	x	x
Teso district, Uganda	Transport	x	x
Embu district, Kenya	Recent migration	x	x
Machakos district, Kenya	Recent migration	x	x
Areas around Nairobi, Kenya	Urban demand	—	x

x Mechanized.
— Not mechanized.

But intensification of the agricultural system does not always lead to use of the plow. As mentioned in chapter 2 and discussed further in chapters 4 and 5, location-specific soil or topographic conditions can prevent the transition from hand power. Even where the plow has not been incorporated into permanent systems of cultivation, however, the transition from human labor is sometimes made for selected power-intensive operations such as transport and milling. In the Kenyan highlands, for example, motorized transport is used to carry goods to the market. Bicycles and motorcycles have replaced head porterage in most of Sub-Saharan Africa. Similarly, in virtually every corner of Sub-Saharan Africa hand milling has been replaced by diesel or electric mills. In chapter 9 further discussion of the mechanization of milling and transport will be found.

Table 3-3 shows the operations mechanized in the permanent systems of cultivation cited in this chapter. Transport is consistently the first operation to be mechanized as the society moves to intensify its systems of cultivation, while use of the plow does not always emerge.

Chapter 4

Other Factors That Affect the Profitability of Mechanical Inputs

Intensification of the agricultural system is a necessary although not a sufficient condition for the substitution of animal or mechanical power for human power. The switch to the plow is made only when it is profitable to do so. The profitability of substituting the plow for the hand hoe is determined by the ease with which the soils can be worked, the location on the toposequence—the slope of land on which agricultural activities are carried out—where the farming activity is concentrated, the crops grown and their response to tillage, and the potential capacity utilization of the equipment.

Determinants of the Yield Response to Tillage

The three primary determinants of the yield response to tillage are the type of soil, the location of the farming activity on the toposequence, and the crop-specific response to tillage. These factors explain many of the intraregional differences in the profitability of tillage.

Soil Type

Tillage improves the physical condition of the soil, especially of the superficial layer, by increasing total porosity and changing the pore-size distribution, thus improving aeration, root penetration, and water infiltration and reducing evaporation (Nicou and Charreau 1980, 373). The importance of tillage increases as porosity decreases: it is unnecessary to till coarse sands, absolutely essential to till clays. In addition, tillage may significantly help control weeds in all types of soils. Table 4-1 summarizes the need for land preparation, by soil type.

Table 4-1. *Soil Characteristics and the Benefits of Plowing*

Type of Soil	*Waterlogging and waterholding capacity*	*Risk of drought*	*Difficulty of land preparation*	*Effect of land preparation on yield*
Clay	High	Low	High	High
Clay loam				High
Silt loam				Intermediate to high
	to	to	to	
Sandy loam				Intermediate to high
Loamy sand				Low
Sand	Low	High	Low	Nil

Clays have extremely low porosity and are impermeable to air and water. Tillage is therefore necessary to improve water infiltration and root penetration. The yield response to tillage is therefore very high on these soils. In the absence of an animal or motorized source of power, they are virtually impossible to cultivate and are therefore left for grazing.

Loamy soils are generally well drained and have good structure, especially as the clay content decreases, and they are therefore easier to work by hand. The level of land preparation required is related to the content of clay and organic matter. Clay loams and silt loams with high clay content have drainage problems and therefore require tillage. But even loamy soils containing less clay—sandy loams and loamy sands—tend to develop low porosity and poor rooting conditions, because intensive cultivation leads to a sharp decline in organic matter content.[1] Tillage to a moderate depth, with incorporation of organic matter such as crop residues, green manure, or straw, can alleviate the problem of low porosity and compensate to some extent for the loss in organic matter (Nicou and Charreau 1971). This rate of decline in organic matter content with continuous cultivation tends to increase with temperature and decrease with rainfall and altitude.[2] The yield response to tillage on loamy soils therefore increases with clay content and with farming intensity. Finally, coarse sands are very easy to cultivate by hand but their water-retention capacity is very low and the risk of drought is high.

The ideal soils are sandy loams that contain much organic matter. These soils have good structure and high moisture capacity, and they can be worked fairly easily and rapidly after the rains. Since they have considerable moisture capacity, they can be tilled for a considerable time after the rains stop. The incidence of such soil conditions in Sub-Saharan Africa, however, is limited mainly to the highlands.[3]

Toposequence and Soil Types

In a movement from the summit of the hill to the floor of the valley, changes in soils, usually in depth and clay content, are typically to be observed. This succession of soils down a slope is known as a catena. On the top of a ridge the soil tends to be shallow and rocky. At midslope it is of moderate depth, while at the bottom of the valley it is generally deep and has a high clay content. Consequently, the water-retention capacity of the soil improves as one goes down the slope. There are differences in drainage as well. The soils on the upper slopes are well drained, while those in the valley bottom are often waterlogged.

Differences in soil type across a toposequence as characterized here could be microvariations limited to a few hundred meters or a few kilometers, or where entire regions are part of one level of the toposequence, they could be macrovariations. Most of the land in the northeastern part of Thailand, for example, exhibits the characteristics of the upper slopes, while the central plain is a large low-lying area.

Figure 4-1 shows differences in type of soil by toposequence. The relatively light soils on the upper slopes are easy to work by hand, while the power required for tillage is the greatest in the valley bottoms because of the high clay content of the soil. Also, the bottomlands are virtually impossible to cultivate in the absence of investments in water control and drainage. On the other hand, because water-retention capacity increases with movement down the slope, the risk of drought is greatest on the upper slopes and least in the valley bottom. In his choice of a place to cultivate, the farmer therefore trades off between lesser labor requirements for preparation of the land and greater risk of drought.

The trade-off between greater input of labor and risk of a lower yield varies with agroclimatic zone, becoming less severe as humidity increases. Table 4-2 presents soil preferences by agroclimatic zone and population density. Under arid conditions lower slopes and depressions are the only lands cultivated because only here is the water-holding capacity sufficient to allow any cropping at all. This is the reason intensive cultivation systems of an oasis type are found in arid regions, even at low population density. Pockets of arid-land farming in pastoral areas of northwestern Botswana exemplify this phenomenon.

Figure 4-1. *Toposequence and Soil Type*

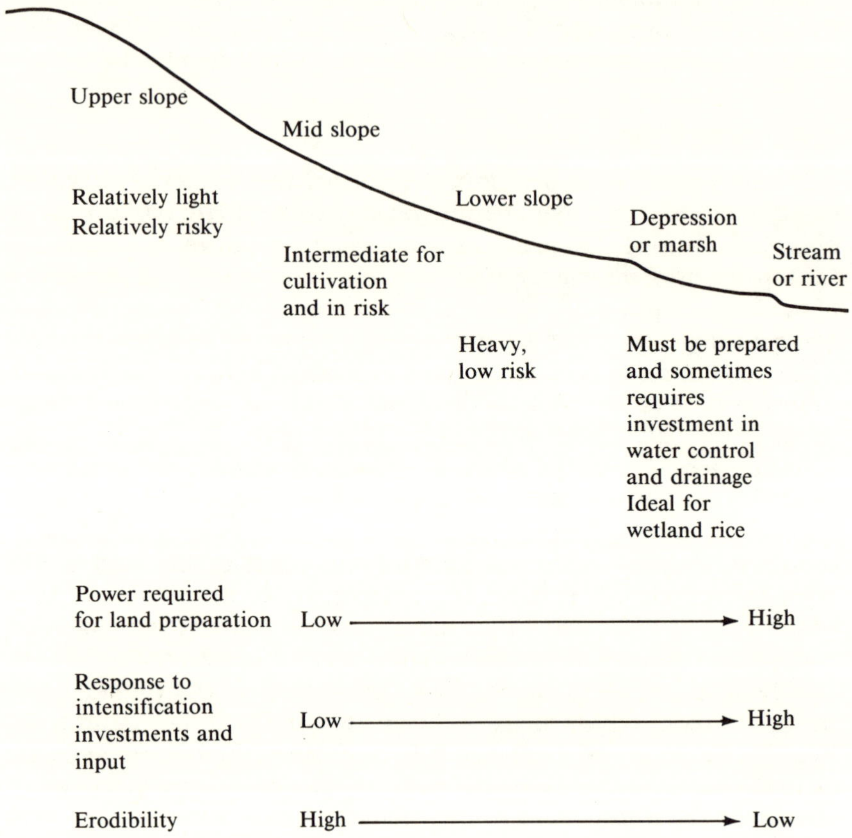

Under semiarid conditions the mid slopes are the first to be cultivated. As population density increases cultivation replaces grazing on the lower slopes and eventually in the depressions. The returns on investments in labor for drainage and irrigation in the bottomlands increase, especially for rice cultivation. The digging and maintenance of drainage facilities is extremely labor-intensive and often avoided until population pressure makes the cultivation of this land remunerative. Also, as population density increases, the supply of labor increases, making it possible to undertake the required investments in labor. Sources of power for tillage are first used in the bottomlands around the time when population pressure makes these lands valuable for cultivation. In many semiarid areas of Africa where hand cultivation is

Table 4-2. *Farming Intensity, Agroclimate, and Preferred Location of Cultivation*

Agro-climate	*Forest-fallow and bush-fallow*	*Grass-fallow*	*Permanent cultivation*
Arid	Lower slopes and depressions only	Lower slopes and depressions only	Lower slopes and depressions only
Semiarid	Mid slopes	Plus lower and upper slopes	Plus depressions
Subhumid	Mid and upper slopes	Plus lower slopes	Plus depressions
Humid	Mid and upper slopes	Plus lower slopes	Plus depressions

predominant, the lower slopes and depressions are still left for grazing, while in the semiarid zones of India virtually all the depressions are intensively cultivated, using elaborate systems of irrigation and animal traction.

For all but the arid areas the foregoing discussion can be summarized as follows: as population density increases, societies have three choices: to intensify production on the mid to upper slopes; to move to more marginal lands in the upper slopes; or to move down the slope. All these choices imply investments either in erosion control or drainage and water control. Typically, the return on these investments is highest on the lower slopes and the valley bottoms. Where lower-slope and valley bottomland are available, therefore, population growth leads to intensification investments in these lands.

On many tropical soils, intensification on the mid slopes will lead to high levels of erosion. The exceptions are soils with high Cation exchange capacity (CEC), such as volcanic soil, alluvial soil, or vertisol (Kang and Juo 1981). When little land is available in the lower slopes, intensification on the mid slopes is associated with investments in bunding, ridging, and terracing.

Except on the high-CEC soils, the returns on intensification investments are lowest on the soils of the upper slopes. As long as lower- and mid-slope land is available, therefore, the upper slopes will not be cultivated intensively. When the better lands become scarce, a farming system with low levels of input and low yields appears on the upper slopes. Because investments in erosion control bring fairly low returns, they may not be undertaken and these marginal lands may degrade. Intensification in the subhumid and humid tropics may also be inhibited by leaching and acidification of the soil. The exceptions, again, are the high-CEC soils. The technological choices available for the subhumid and humid tropical areas susceptible to leaching problems will be discussed in chapter 10.[4] Suffice it to say here that intensive production of field crops with mechanical equipment is not possible on soils in danger of such degradation.

The exploitation of the depressions and valley bottoms in Africa for food production, a more recent phenomenon than similar agricultural production in South Asia and Southeast Asia, can be expected to be critical to the future food supply of the humid tropics. "A close similarity exists between the hydromorphic soils of West Africa with those in Sri Lanka. The long use of these soils for successful rice production in Sri Lanka suggests that considerable potential exists for their use in West Africa" (Greenland 1979). Permanent cultivation of the bottomlands is not subject to the same problems of soil degradation as that in the uplands. It requires substantial investments in water control and drainage, however, and high levels of power for land preparation.

From the foregoing discussion it can be seen that population pressure leads to a sharp reversal in preference—and consequently of price—of different types of land in all but the arid zones. As population density increases there is a shift to land that requires substantially greater input of labor but on which the marginal return on labor and other investments does not decline rapidly. The consequences of these land preferences is that the shift to the plow is most likely when societies start cultivating the clays of the lower slopes and valley bottoms. The use of the plow typically spreads to the mid and upper slopes only after its use in the bottomlands has been established.

Table 4-3 presents several examples of land use across the toposequence for several areas in Sub-Saharan Africa that are consistent with our thesis. The systematic nature of this evolution of soil-use patterns, however, became clear to us from observations made during many of our field visits. We therefore do not have data on this evolution for all the locations mentioned in this study.

Crop-Specific Response to Tillage

Between 1952 and 1969, the Institut de Recherches Agronomiques Tropicales (IRAT) conducted 250 experiments in French-speaking West Africa, comparing unplowed control areas and plowed areas. On the basis of these experiments it was concluded that in dry tropical areas the incremental yields vary from crop to crop, being highest for rainfed rice and lowest for groundnut (Charreau 1974, 237). Summary results of IRAT experiments are presented in table 4-4. Tillage leads to the

Table 4-3. *Toposequence and Land Use in Selected Regions of Africa*

Country or region	*Agro-climatic zone*	*Population density*	*Pattern of land use*			
			Upper slopes	*Mid slopes*	*Lower slopes*	*Depressions*
The Gambia[a]	Subhumid	Low	Forest	Rainfed groundnut, millet, cotton	Palm forest	Flooded rice
Ukara Island, Tanzania[b]	Subhumid	Very high	Grazing	Rainfed millet, manioc	Legumes	Irrigated rice, sorghum, sweet potato, vegetables
Sukumaland, Tanzania[c]	Subhumid	High	Grazing in wet season	Cassava, cotton, legumes	Sorghum, maize	Irrigated rice, sweet potato, use of the plow
Mazabuka, Zambia[d]	Semiarid	Moderate	Grazing	Maize, use of the plow	Maize, use of the plow	Uncultivated
Zambesi basin, Zambia[d]	Semiarid	Very low	Forest	Maize, root crops	Maize, taro	Uncultivated
Ngamiland, Botswana[c]	Arid	Very low	Grazing	Grazing	Millet, sorghum	Sorghum, maize, use of the plow

a. Ruthenberg (1980).
b. Ludwig (1968).
c. Rotenham (1968).
d. Trapnell and Clothier (1937).
e. Schapera (1943).

Table 4-4. *Mean Effect of Plowing on Yields of Crops in West Africa*

Crop	*Number of experimental observations*			*Yield on unplowed controls*	*Increases brought about by plowing*		*Increase in revenue*
	Total	*Beneficial*	*Percent beneficial*	*(kilograms per hectare)*	*Kilograms per hectare*	*Percent*	*(U.S. dollars per hectare)*
Rainfed rice (paddy)	11	11	100	966	549	36	556.50
Sorghum (grain)	46	29	85	1,874	536	29	47.43
Maize (grain)	6	6	100	2,093	568	27	53.75
Cotton (grain)	7	7	100	1,629	433	27	779.40
Pearl millet (grain)	22	21	95	1,245	256	21	n.a.
Groundnut (pods)	31	27	87	1,412	274	19	119.17

n.a. Not available.

Note: Vegetation was removed before plowing. These are average yield effects of experiments conducted by the Institut de Recherches Agronomiques Tropicales (IRAT) between 1970 and 1979 on sandy to coarse loamy soils in semiarid West Africa. Revenues were calculated using average world prices, 1970–79, for each crop.

Source: Nicou and Charreau (1971).

greatest increase in yield in rainfed cultivation of rice; on the average, in all experiments, the rice yields rise from 966 kilograms to 1,515 kilograms a hectare, or 36 percent. All other increases in yield are small in comparison to the incremental rice yield. The second highest yield increase, 29 percent, is in sorghum, followed by maize and cotton at 27 percent.

We converted these absolute yield increases into revenue terms using the average world price, 1970–79, for each crop. The ordering by incremental revenue is substantially different from the ordering by increases in quantity. Plowed cotton fields bring the highest incremental revenue per hectare, followed by rice, then groundnut. Average incremental revenue for the three crops is US$780, $556, and $120, respectively. Incremental revenues for maize and sorghum are only US$54 and $47, respectively.

Using this information, it is possible to conclude for semiarid Africa that where transport facilities and trade policies allow international prices to be transmitted to agricultural areas, the transition to the plow is most profitable where cultivation of cotton, rice, groundnuts, or maize, in that order, is common. Several of the instances mentioned in chapters 2 and 3 show a strong association between use of the plow and the growing of these crops.

The foregoing conclusion would probably hold for the subhumid tropical areas also, although we have no information on tillage experiments in this zone. Rice will typically be the most important crop in the humid tropical depressions that are brought under the plow. As for the upland fields in the humid tropics, the conclusion cannot be extended, since the viability of permanent production of field crops and use of the plow in this zone has not been established (see chapter 11).

The Effect of Season Length on Potential Utilization of Capacity

Yield response to tillage, by type of soil and by crop, determines tillage requirements and whether the plow ought to be used in a given area. For given yield levels, however, the profitability of using the plow is also related to the potential capacity utilization of equipment, which in turn depends primarily on the number of days available for land preparation. Severe constraints on capacity utilization can prevent the adoption of the plow, even where the effect on yield is fairly high. Conversely, where the plowing season is long, use of the plow may be profitable even with only a modest effect on the yield.

The number of days available for land preparation is the greatest in humid tropical climates and the least in the arid zones. Within a particular climatic zone the length of the growing season increases with altitude and with the moisture-holding capacity of the soil. Table 4-5 shows the relation between agroclimatic zone and the length of the growing season.

During the dry season evaporation is high, and in the absence of rainfall, reserves of soil water are used up rapidly and the vegetation tends to dry up. No cultivation can take place because soils are too dry—and in West Africa too compact—to be worked without damage to implements. In much of arid and semiarid Africa soil preparation must therefore wait until the onset of the rainy season, when enough rain has been received to wet the soil to the depth of plowing. In these areas, however, delayed planting often causes a reduction in the effective length of the growing season and sharply decreases crop yields.

> A precise timetable of farming operations imposed by the water regime is of paramount importance for the timely and successful preparation of land and for the establishment of crops, weeding, the efficient use of fertilizers and for productivity. Delayed farming operations not only adversely affect subsequent operations (weeding, spraying of crops) but result in a reduction of the growing period and an increased risk of water stress during the period of maturity leading to reduced yields (Kowal and Kassam 1978, 94).

Farmers therefore face a trade-off between plowing and sowing that can be described as follows: while tillage can improve the structure of the soil and therefore increase yields, its inevitable result is delays in sowing. But delayed sowing means decreased yields. Plowing is advan-

Table 4-5. *Potential for Animal-drawn Plows, by Agroclimatic Zone*

Climatic zone	*Average annual rainfall (millimeters)*	*Length of growing period (days)*	*Remarks*
Arid	0–400	<75	Plowing-sowing trade-off severe. Use of plow confined to heavy soils where absolutely necessary.
Semiarid			
Arid fringe areas	400–500	75–90	
Low to medium rainfall zone	500–650	90–120	Use of plow quite common.
High rainfall	650–800	120–150	
Subhumid	800–1200	150–270	Use of plow faces constraints other than capacity utilization.
Humid	>1200	>270	

tageous only where its effect on the yield exceeds the losses in yield associated with delays in sowing. The plowing-sowing trade-off is more severe in areas in which rainfall is low and on soils that have low moisture-holding capacity. Consider the extreme case of the arid fringe areas in which annual rainfall is less than 500 millimeters. The growing period is shorter than 90 days, so the result of delayed sowing is sharply higher risks of crop loss. Light sandy soils can be prepared fairly easily by hand, while plowing affects the yield in several ways. Under these circumstances the farmer is better off sowing directly, without plowing, a larger area than he could prepare in time with the plow.

Even under arid conditions, however, the heavy clay soils in the depressions and valley bottoms must be plowed if they are to be cultivated at all. Their higher water-holding capacity, moreover, lengthens the growing period and thus the time available for plowing. Accordingly, in arid regions with a high proportion of such soils the plow is used for land preparation. The depressions in Kordofan, Sudan, for example, are intensively cultivated with draft animals, while the surrounding upland areas are seeded directly by means of a hoe, without preparatory tillage (McCown and others 1979). We observed similar patterns of plow use in northern Burkina Faso and Niger.

It is also appropriate at this point to mention one exceptional use of draft power in the arid fringe areas, the use of horses to pull groundnut seeders in Western Senegal. This instance is exceptional because in this area, where the growing season is short and the soil is light and sandy, seeding was the first operation to be mechanized. The probability is high that groundnuts sown into the moist, sandy soil will germinate. The period during which the soil is moist is short, however, so there is a premium on speedy completion of the operation. The use of the seeder allows the cultivation of larger areas within the time available. Horses are used rather than oxen because they are faster.

Unless soils are heavy, the usefulness of animal- or tractor-drawn plows is limited in much of the semiarid tropics where rainfall is low, since the farmer cannot accrue the benefits of spreading equipment costs over a larger area, among several users, or both. The potential for the use of the plow is highest in the semiarid zones, where rainfall is high. Because of the greater length of the plowing season, the trade-off between plowing and sowing is a lesser constraint. A farmer can achieve a high capacity utilization of equipment either by cultivating a large area himself or by renting out his equipment to other farmers. That it is in the high-rainfall semiarid tropics that the use of the plow is most widespread is consistent with this analysis.

From the point of view of capacity utilization there are few constraints to plowing in the subhumid and humid lowlands and in the tropical highlands. In the subhumid and humid lowlands adoption of the plow has been confined primarily to the cultivation of rice in depressions and valley bottoms, for reasons discussed earlier. In upland fields, however, the hand hoe has been tenacious, despite repeated attempts to introduce new sources of power. This is mainly because of the constraints on intensive production of field crops associated with leaching and erosion. Techniques of permanent cultivation of these soils involve dense mixtures of plants of different duration, so the use of animal-drawn plows is infeasible or unnecessary.

Utilization of Capacity through Rental Markets

Opportunities for increasing capacity utilization of equipment through rental markets for time-bound and synchronic operations are limited because of sharp conflicts about timing among potential users. This conflict arises because it is impossible to decide in advance the best sequence in which to serve the various customers of the same rental service. It is therefore no accident that few rental markets exist for seeding, which is a highly time-bound operation, even in climates with long growing seasons. Rental markets for grain-harvesting equipment are rare for similar reasons. Rental markets for threshing, milling, and transport equipment are common, because these operations can typically be stretched over several months without affecting output levels at all.

Rental markets for land-preparation equipment are more likely to be established in areas where the plowing operation is not severely time-bound—where dry-season or postharvest plowing is possible or where delayed plowing and planting do not lead to sharp declines in yields. Land preparation is not severely time-bound in most temperate zones, for example, where temperature, rather than soil moisture, limits the growing season. Several months are available for land preparation in Western Europe, where plowing is done both after harvest in the fall and before sowing in the spring. But the land preparation period in the lower-rainfall semiarid zones is very short and there is sharp conflict about the timing of plowing. Rental markets cannot therefore be expected to solve the capacity utilization problem in these regions.

Capacity utilization of tractors can sometimes be increased by providing contract-hire operations across several rainfall zones. Successful instances of such migrant operations can be observed in Thailand and

India. Migration of contract-hire operators of tractors is also to be seen in Nakuru and Narok, Kenya, and will probably emerge in other parts of Sub-Saharan Africa. Similar migration using oxen and plows is typically infeasible because of the high cost of moving animals.

The other means of overcoming low capacity utilization usually advocated is to use the animal or tractor for other agricultural operations. The threshing operation is not time-bound, and tractor-powered threshers are a possibility. Tractors, however, are inefficient sources of stationary power, and in the Indian Punjab, the Philippines, and Central Thailand stationary machines usually have their own sources of power. Most of these machines are used under rental systems. The same pattern could be expected to be followed in Africa as harvested volumes increase and wages rise.

Transport is usually not time-bound either and can easily be provided by a contract-hire service. The demand for animal-drawn carts or tractor-drawn trailers depends on the marketed quantities, the amount of organic manure required and the availability of alternative means of transport. Where marketed quantities are large or where high levels of manure are used to maintain the fertility of the soil the demand for transport services provided on hire will be great. The demand for oxcarts or tractor trailers is also higher in areas where alternative transport services, such as trucks, lorries, bush taxis, and buses, are costly or absent.

High-Potential Tropical Highland Areas That Have Bypassed the Plow

The tropical highlands are the most densely populated and intensively cultivated parts of Sub-Saharan Africa. Tropical highlands are often characterized by long growing periods and good soil structure associated with a high content of organic matter. High intensities of land use are usually complemented with terracing, manuring, and mulching. In many of the tropical highlands, however, the use of the hand hoe has persisted.

Densely populated highlands under intensive production of field crops have indeed switched to the plow. The central highlands of Ethiopia have historically been densely populated and have intensively cultivated the field crop teff. The transition from the hand hoe to the plow took place in Ethiopia several centuries ago. More recently, intensification of farming in certain densely populated pockets of the Kenyan highlands, such as Nyanza province and the Machakos district, has led to the use of animal-drawn plows in the production of maize.

Use of the plow on the highlands has not been widespread either because slope and stones make it difficult or because these locations favor tree crops and the production of milk and meat.

Where soils are not susceptible to erosion and cultivation is feasible on steep slopes, the slope makes it physically difficult, if not impossible, to use the plow. Stones and rocks in a field present the same constraint to the plow as tree stumps. Neither prevents annual cultivation, however. Agriculture is practiced on very steep slopes in Rwanda, Burundi, and the Kigezi district of Uganda. Small terraced plots are intercropped with field and root crops, and human labor is the only source of power used. For further details on highland farming systems see Jones and Egli (1984), Ruthenberg (1980), and Morgan (1969).

The Kikuyu districts of Kenya illustrate the comparative advantage of the highlands in the production of tree crops such as tea and coffee, and of vegetables, milk, and meat rather than field crops. Field crops such as maize and sorghum are grown on small garden plots mainly for home consumption; hand hoes are used to till these plots because the power requirements are relatively low and the period available for land preparation is long.

Notes

1. This downward trend in organic matter content is halted and is succeeded by an upward trend during the fallow period, because vegetative growth supplies the soil with fresh organic matter. The original levels of organic matter can be reached within three to ten years of fallow (Ahn 1970, 238).

2. As rainfall decreases, a distinct and prolonged dry season emerges, leading to a more rapid decomposition and hence lesser net accumulation. An increase in altitude is accompanied by an increase in rainfall and a decrease in temperature, both of which lead to increases in accumulation and a decline in decomposition of organic matter.

3. For detailed descriptions of the soils of Sub-Saharan Africa, see Ahn (1970), Greenland (1979), Taylor (1980), Sanchez (1976), and Webster and Wilson (1980).

4. "Production of food crops (in the humid tropics) is only successful when small areas are cleared and cropped with a mixture of species which maintain a continuous vegetative cover over the soil, until the land reverts back to fallow" (Greenland 1979). As population density increases, fallow periods decrease and cultivation takes place on small plots that are intercropped with tree and field crops.

Chapter 5

Animal Draft and Tractor Power in Sub-Saharan Africa: A Historical Overview

While the use of draft animals in Asia and Europe dates from several millenia before Christ, their use in Sub-Saharan Africa is barely a century old. In fact, the use of draft animals in several parts of Sub-Saharan Africa began as recently as the 1920s and 1930s. Before the twentieth century the use of animals for draft was limited mainly to the carrying of loads by donkeys and mules and the carrying of people by mules, horses, and camels. Ethiopia is the single exception in all Sub-Saharan Africa where the use of the scratch plow for tillage is historic. According to the world classification of scratch plows established by A. G. Haudricourt and Jean Delamare (1955), the scratch plow most commonly found in Ethiopia was the *chambige,* a beam with a curved section. This type was typical of historic Northern Europe, where its use continued even after the advent of the moldboard plow. It was used until recently in the Auvergne, in central France, where it received its name because of the arched shape of the beam. The chambige is still used in areas of Spain, Turkey, Syria, and Lebanon and in other widely scattered areas throughout the world. It should be emphasized that the domestication of animals was as ancient in Africa as in other parts of the world, but they were not used for draft purposes.

When animal-draft power began to emerge in Sub-Saharan Africa it followed a pattern similar to that followed in other parts of the world. The first operation to be transferred from human to animal power was usually transport, and only then came plowing. The history of animal-drawn plows in Sub-Saharan Africa falls into three distinct periods: the colonial period, before 1945; the postwar, preindependence period, 1945–60; and the postindependence period. In each period certain conditions determined the successful transition to the plow. Table 5-1 presents several examples of the spontaneous spread of animal traction in Sub-Saharan Africa.

Table 5-1. *Period of Introduction and Early 1980s Use of Animal Traction in Selected Areas of Sub-Saharan Africa*

		During period of introduction			*In early 1980s*			
Period of introduction	*Country and region in which introduced*	*Dominant draft animal*	*Primary crop*	*Principal implement*	*Dominant draft animal*	*Primary crop*	*Principal implement*	*Number of draft animals in early 1980s, by country*
?	High plateaus, Ethiopia	n.a.	n.a.	Scratch plow	n.a.	n.a.	n.a.	n.a.
1880–90	High plateaus, Madagascar	Cattle	Rice	Plow	Cattle	Rice	Plow	200,000
	Botswana	Cattle	Sorghum	Plow	Cattle	Rice	Plow	n.a.
	Mozambique	Cattle	n.a.	Plow	Cattle	n.a.	Plow	10,000+
1920–30	Central Guinea	Cattle	Rice	Plow	Cattle	Rice	Plow	100,000
	Northern Nigeria	Cattle	Groundnut	Ridger	Cattle	n.a.	n.a.	n.a.
1930–40	Mali river bottoms	Cattle	Rice	Plow	Cattle	Cotton	Plow	230,000
	North central Senegal	Horse	Groundnut	Seeder	Horse	Groundnut	Seeder	480,000
	Northern Ghana	Cattle	Groundnut	Ridger	Cattle	n.a.	n.a.	20,000
1950–60	Northern Cameroon	Cattle	Cotton	Plow	Cattle	Cotton	Plow	40,000
	Central eastern Gambia	Cattle	Groundnut	Ridger	Cattle	Groundnut	Multipurpose cultivator	30,000+
	Southern Niger	Cattle	Cotton	Scratch plow	Cattle	Various	Multipurpose cultivator	35,000
1960–70	Southern Chad	Cattle	Cotton	Plow	Cattle	Cotton	Plow	105,000
	Central Burkina Faso	Donkey	Groundnut	Scratch plow	Cattle	Cotton	Plow	140,000
	Northern Benin	Cattle	Cotton	Plow	Cattle	Cotton	Plow	25,000
1970–80	Northern Côte d'Ivoire	Cattle	Cotton	Multipurpose cultivator	Cattle	Cotton	Multipurpose cultivator	30,000

n.a. Not available.

Pockets into Which Animal Traction Spread Spontaneously before 1945

Colonization of Africa and the establishment of settler farms began in the nineteenth century. Settler farmers were often established in cooler areas, where the soil was predominantly sandy loam, which is particularly well suited to plow-based agriculture, and where they could replicate the short-fallow or permanent farming system of Europe. African farmers were often expelled from these areas to regions where conditions for use of the plow were less favorable. The markets for export crops were frequently reserved for settlers, moreover, depriving the local population of incentives for extending the area under cultivation.

Provided that clearing and destumping the land was not too difficult and there were cattle in the area, settlers immediately began to use oxen for traction. The use of oxen on settler farms was therefore common in all but the very arid regions and the humid lowlands. The spread of animal traction from the settler farms to the local population was limited but in selected pockets proceeded rapidly. Instances of spontaneous spread of animal traction are distributed widely from Madagascar and southern Botswana to northern Nigeria and central Senegal. The conditions that determined its successful adoption, however, were varied.

The earliest instance of the spontaneous spread of animal-draft power was around 1880 in Madagascar, where the situation was atypical of the rest of Sub-Saharan Africa. The economy was based on paddy rice cultivation, the area was densely populated and devoid of tree stumps, and the animals were large enough to use European equipment (Labrousse 1961). The association between wet rice cultivation in the lowlands and the spread of ox-drawn plows is also clear in Guinea around 1925 and in Mali and northern Sierra Leone after 1930. In all these locations the moldboard plow was about the only animal-drawn implement adopted. The use of the plow spread to the uplands only after it had become well established for wet rice cultivation in the lowlands.

Botswana followed Madagascar in adopting the plow. The use of a moldboard plow pulled by a team of two to eight oxen became established in southern and eastern Botswana as early as the turn of the century. Animal traction in southern Botswana had been introduced by the Boer farmers during their famous *Trek* (1834–38) from the British (Hailey 1957). The transition to the plow in Botswana was encouraged because raising livestock is the primary activity, crop production is concentrated in the lowlands and depressions, where water stress is not severe, and soils are heavy. Accordingly, when animal-drawn plows were introduced, animals were already available and most of the fields were stump-free. The benefits of using the plow, moreover, were considerable.

Between 1925 and 1930 animal traction spread to northern Nigeria and to northern and central Senegal. In both cases the motivating force was the high export demand for groundnuts, but the techniques adopted were completely different. In Senegal the horse-drawn seeder was the only implement adopted, and in Nigeria it was an ox-drawn ridger. The plow was not adopted in either country. In chapter 3 we discussed the function of the railway in promoting intensive production of groundnuts in the two countries. It would be instructive, however, to understand why such different tools were chosen in the two areas. As discussed in chapter 4, the light sandy soils in Senegal do not require tillage and can be seeded directly when moist. The speedy completion of seeding is essential in order to ensure germination, however, something made possible on a large area by the use of horse-drawn seeders. Groundnut growing areas in Northern Nigeria were concentrated on the sandy loams of the middle slopes. These soils are susceptible to high levels of erosion, which can be controlled by ridging, a labor-intensive operation if done with the hoe, hence the rapid adoption of the ox-drawn ridger.

On the periphery of Lake Victoria, in Nyanza province, Kenya, Teso district, Uganda, and Sukumaland, Tanzania, agricultural production, especially of cotton and maize, expanded rapidly during the 1920s. In Kenya and Uganda agricultural intensification was induced by the railway that connected Kisumu to Mombasa, and in Sukumaland a road connection had a similar effect. Animal-drawn moldboard plows became common in all these areas in the 1920s and 1930s (see table 5-2). Teso district, for example, which lies in the basin of Lake Kyoga, has traditionally been a densely populated area of Uganda, with a population of thirty persons a square kilometer as early as 1920. The more densely populated southern part of the district had as many as fifty persons a square kilometer in 1920. Accordingly, at that time the southern counties were under grass-fallow while those in the north were still under bush-fallow. The ox-drawn plow and cotton were introduced at approximately the same time, around 1910. Cotton spread rapidly throughout the district while the plow spread mainly in the southern counties. Table 5-3 shows the spread of cotton and the plow in Teso district, where the number of plows rose from 282 in 1923–24 to 15,388 in 1936–37; cotton acreage rose during the same period from 84,000 to 158,878. Note that despite equal effort to promote use of the plow in

Table 5-2. *The Spread of Animal-Draft Power in Selected East African Locations around Lake Victoria*

Year	*Teso district, Uganda*	*Sukumaland, Tanzania*	*South Nyanza province, Kenya*
1923–24	282	n.a.	n.a.
1929–30	6,170	n.a.	n.a.
1936–37	15,388	n.a.	n.a.
1939–40	n.a.	Few	n.a.
1943	n.a.	n.a.	2,081
1952	n.a.	5,000	n.a.
1960	40,000	n.a.	n.a.
1961	n.a.	20,000	n.a.
1981–82	n.a.	50,000	52,000

n.a. Not available.

Sources: Uganda, Tothill (1940), Uchendu and Anthony (1975a); *Tanzania,* Iliffe (1979) and Kjoerby (1983); *Kenya,* Humphrey (1945) and Pollard (1983).

Table 5-3. *The Spread of Cotton Cultivation and Use of the Plow in Teso District, Uganda, 1922–37*

Season[a]	*Number of plows in use*	*Yield of Cotton (pounds per acre)*	*Cotton acreage*
1922–23	?	314	84,000
1923–24	282	307	68,000
1924–25	734	360	87,500
1925–26	1,154	287	97,500
1926–27	2,710	202	99,329
1927–28	2,941	250	69,737
1928–29	3,400	218	132,000
1929–30	6,170	60	117,265
1930–31	6,423	329	118,813
1931–32	7,849	185	115,626
1932–33	8,280	212	134,481
1933–34	9,913	225	112,222
1934–35	11,615	119	112,329
1935–36	13,726	311	116,812
1936–37	15,388	n.a.	158,878
1940	24,602	n.a.	150,800
1945	25,279	n.a.	183,300
1950	n.a.	n.a.	176,200
1955	40,000	n.a.	265,300
1960	40,000	n.a.	224,400

n.a. Not available.

a. The low yields in 1929–30 were the result of an epidemic of Black Arm disease.

Sources: Through 1936–37, Tothill (1940, 57); 1940–60, Uchendu and Anthony (1975a, 38).

the northern counties and subsidized destumping operations, the plow was not at first adopted there. It only began to be used in these areas in the 1960s, when increasing population density in the northern counties led to a recession in bush cover (Tothill 1940; Uchendu and Anthony 1975a, 1975b; Anthony and others 1979).

These instances are not exhaustive but indicate the general conditions—either the appropriate intensity of farming or the absence of tree stumps in selected pockets of land—that made early uses of animal draft successful. Several early promotion efforts failed, mainly because they were tried when the area was still under forest-fallow or bush-fallow cultivation. Once the appropriate farming intensity has been reached the transition to the plow is then governed by the factors that determine profitability. Examples of attempts to introduce animal traction that were unsuccessful because of inappropriate farming intensities are British efforts in central Uganda and northeastern Zambia, the German experience in Arusha province of Tanzania, French projects in Guinea, and Belgian efforts in Zaïre.

Colonial attempts to introduce animal traction earlier than 1945 suffered from some of the following problems:

- Poorly developed veterinary services made the ownership and use of animals costly, especially in the sparsely populated lowlands. The tsetse fly, for instance, prevented the spread of animal-draft power in Southern Nigeria, Southern Guinea, parts of Uganda, and Tanzania, among other places.
- Imported farm equipment was often not suited to specific local requirements. The first plows were too heavy for African oxen and for use in partially cleared fields. Early studies showed a distinct preference for lightweight but solid plows and ridgers, which required lower tractive power and could be lifted over stumps. The refinement of equipment for local conditions began with lightweight hoes and seeders in Senegal in the late 1920s. Manufacturing industries responded by producing plows, ridgers, seeders, and hoes designed especially for African conditions.
- Early attempts to introduce animal traction were often associated with a whole series of changes in farming techniques, which tended to hamper rather than promote the use of draft power. Farmers being taught to plow were simultaneously being taught to apply fertilizer, make manure, incorporate crop residues, or introduce complex crop rotation, and they generally did not clearly understand the technical and financial connections among the various techniques. According to our analysis in chapter 2, many of these

yield-increasing techniques were far too labor-intensive to be attractive when land was abundant, as it usually was at that time.

- The spread of the plow from settlers to smallholders was inhibited by a number of discriminatory practices, which explain why it was limited to a few pockets. We have already shown that access to markets is an important driving force for intensification and thus for the use of the plow. But where settlers were present, as in Kenya, the production of export crops by smallholders was prohibited as part of a policy to encourage them to work on settlers' farms. Labor constraints on these farms were far more compelling than on the smallholders' farms, on which subsistence requirements could be produced easily without expansion into larger areas. Alienation of land for settlers, moreover, was concentrated systematically where the soils were most suitable for European farming with the plow. In the small regions, such as the Mazabuka district of Zambia, where access to markets and the quality of the soil were as good as on European farms, use of the plow did indeed spread rapidly (Anthony and Uchendu 1970).

The Development of Animal Traction after 1945

The development of animal traction after 1945 is best discussed separately for the preindependence and postindependence periods.

The Preindependence Period

In the immediate postwar period, animal traction was constrained by supply difficulties that were a result of the crash of 1929 and the epidemics that raged among livestock in West Africa, especially after 1935. After the war the biggest constraint to the use of animal traction was the priority given to tractorization. All this led to the virtual stagnation of animal traction until the late 1940s and early 1950s. Postwar efforts, especially in French West Africa, were concentrated on the pilot-farm model (Casse and others, 1965). The pilot farms were a network of farms that were meant to act as models from which the local farmers could learn. These farms, however, typically promoted a host of techniques that were not in demand under local farming conditions. Little thought or attention was given to farming systems or to problems of financing, veterinary services, and spare parts that prevented the adoption of the techniques by small farmers. Sometimes motivated by well-intentioned charitable institutions, these efforts did not bear fruit and could even be held responsible for wasting research funds and misguiding agricultural development projects, including animal-traction projects (Casse and others, 1965; Sargent and others, 1981).

Where its use was feasible and profitable, however, animal traction continued to spread. In Mali, Guinea, and Niger animal-drawn plows began to be used for upland farming and were no longer restricted to wet rice cultivation. The use of animal-drawn plows was also becoming established in the Mazabuka district of Zambia, the Machakos district of Kenya, and the Arusha province of Tanzania. Here again the reasons for its adoption and spread had more to do with agricultural intensification than with any organized extension effort.

Perhaps the most useful contribution of the postwar colonial efforts was research in veterinary medicine and farm mechanization. Advances in veterinary medicine helped control epidemics of rinderpest, cattle plague, and peripneumonia. In Guinea, for example, it was possible to rebuild populations of livestock, including oxen. Veterinary service, in Guinea and elsewhere, became a central part of all efforts to promote animal traction.

The multipurpose tool bar, whose principle is the use of several tools on the same frame, was the most significant contribution of the research efforts in this period. The size of the tool bar varies from a small triangular frame to a large, wheeled tool carrier. Some of the latter are intended also to be used as carts. So far none of the large, wheeled tool carriers has been adopted by farmers to any significant extent.

In French West Africa research was conducted on the development of lightweight frames, for which cultivators, weeders, and ridgers were developed. They performed well on pilot farms but were too expensive for the farmer's field (Haudricourt and Delamare 1955; CEEMAT, 1969). It is the smaller multipurpose tools that have gained widespread adoption, but even these still had to be fine-tuned to accommodate a diversity of local farming conditions. By 1970, however, the multipurpose frame had become more popular than the moldboard plow in French West Africa (see figure 5-1).

Enterprises that manufactured animal-draft equipment were being established throughout Sub-Saharan Africa during this period. By 1961, SISCOMA of Senegal, in cooperation with IRAT, became the first manufacturing firm to undertake the production of a uniform line of animal-drawn implements in West Africa (LeMoigne and Zerbo 1977). The John Holt Company in Nigeria and others followed.

Postindependence Efforts to Promote Animal Traction

The striking feature of postindependence policy was a sudden emphasis on tractorization and a wish to bypass animal traction. The

Figure 5-1. *Changes in the Number of Animal-drawn Implements in Use in French-speaking Western and Central Africa, excluding Guinea and Zaire, 1930–80*

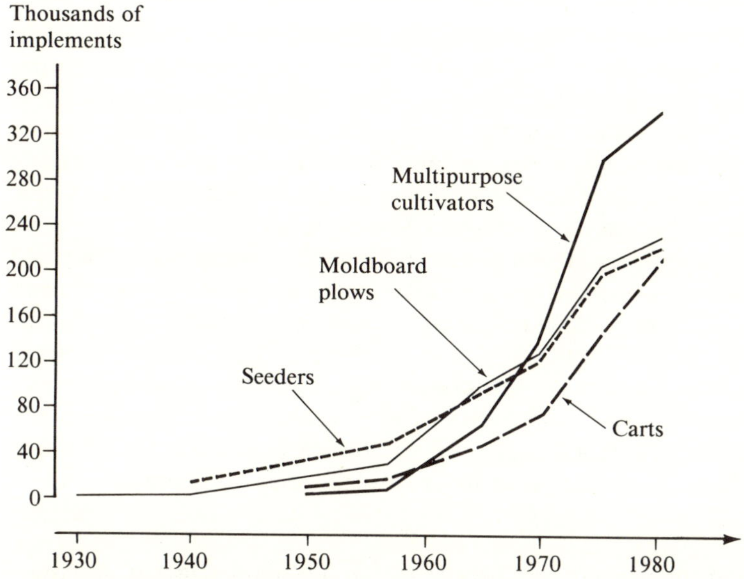

Note: Most seeders are concentrated in north central Senegal; the number of plows would be greater if Guinea were included.

Source: CEEMAT archives and miscellaneous data, by country, after 1960.

majority of government resources for mechanization were used for the purchase and maintenance of tractors, and little was left for promoting animal-draft power. This effort to skip the animal-traction stage was abandoned in the mid to late 1970s, and some efforts were made to reestablish the use of animal-drawn equipment.

The case of Tanzania illustrates this postindependence withdrawal from and subsequent revival of the use of animal power. Tanzania became independent in 1961 and immediately embarked on an attempt to modernize agriculture. Government efforts were turned to promotion of tractor cultivation through mechanized block-cultivation schemes, village settlement schemes, contract-hire schemes, and credit on favorable terms to large private farmers. Apart from the use made of privately owned tractors, the mechanization schemes rapidly turned out to be a fiasco (Kjoerby 1983, 67; for a complete review of the failure of mechanization policy in Tanzania, see I. J. Singh 1977).

A more cautious attitude was taken toward tractor mechanization under the second five-year plan, 1969–74, and greater encouragement was given to ox-plow cultivation and to the design of animal-drawn implements. It was only during the early 1980s that this policy shift began to be implemented in the face of severe foreign-exchange constraints that prevented further imports of tractors. Several other countries—Guinea, Ghana, and Zambia, to name but a few—made similar policy swings. Thus, until the late 1970s discriminatory policies led to a virtual stagnation of animal traction in Sub-Saharan Africa.

Areas in which the plow had been adopted before independence had been achieved continued to use it, often against all odds, as was done in Guinea, but in new areas its adoption was slow because animal-draft equipment was difficult to obtain. Indeed, in East Africa, in areas where animal traction is well established, the plow most commonly seen is the twenty-year-old Victory Plow obtained from British manufacturers before independence. It is still too early to assess the effects of the recent policy turnaround in favor of animal-draft power.

The Introduction of Tractors into Sub-Saharan Africa

The first tractors appeared in Sub-Saharan Africa during the period between the wars. They were used initially on settler farms and government-run farms. After 1945, however, tractors began to be used by African farmers also. Most of the postwar imports of tractors were financed by the fund for farm machinery allocated through the Marshall Plan. Since then bilateral and multilateral aid has been supporting the import of tractors for agricultural purposes. Between 1945 and 1981 there were three significant spurts in the number of tractors in use, with intermediate periods of slow growth (see figure 5-2, table 5-4).

The first spurt came around 1945, and the countries that started to promote the use of tractors during the period 1945–55—Zimbabwe, Kenya, Zambia, and Malawi—can be called the first generation of tractor users. The use of tractors in these countries spread from colonial farms to private farms owned by native Africans. Countries that committed substantial resources to the use of tractors between 1958 and 1970—Tanzania, Ethiopia, Ghana, and Côte d'Ivoire, among others—can be considered the second generation of tractor-using countries (see table 5-4). In many of these countries tractors were provided through cooperative farms, state farms, or tractor-hire services.

Figure 5-2. *The Growth in the Total Number of Tractors in Use in Sub-Saharan Africa, 1947–87*

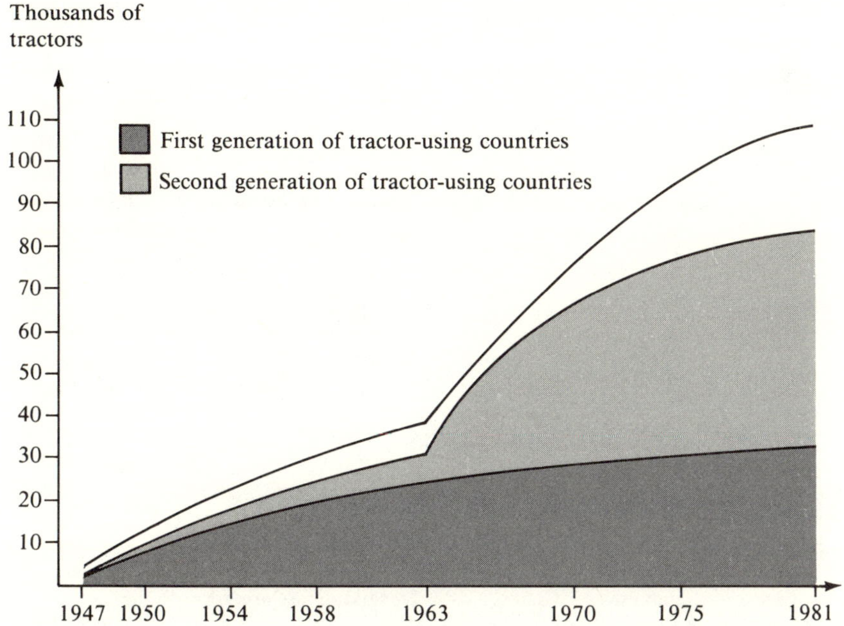

Note: The figure refers to tractors of more than eight horsepower in service in the total population.

Source: FAO yearbooks.

Two more recent spurts can be distinguished. Tractors were purchased with oil receipts in Nigeria between 1970 and 1975, and the number of tractors increased slightly between 1975 and 1981 in some countries, such as Somalia, Zaïre, Uganda, and Cameroon, where they had been fairly rare.

In a number of countries throughout Sub-Saharan Africa, including Burkina Faso, Niger, Rwanda, Burundi, and Liberia, there was never an organized effort to increase the use of tractors, and the number of tractors is very small. Sub-Saharan Africa has, in fact, fewer tractors per thousand inhabitants than either Asia or Latin America. Figure 5-3 shows the number of tractors per thousand inhabitants in Sub-Saharan Africa from 1947 through 1981.

Table 5-4. *Changes in the Rate of Growth in the Number of Tractors in Use in the Principal Countries of Sub-Saharan Africa, 1947–81*
(thousands of tractors in use)

Group of countries	*First generation*[a]	*Second generation*[b]	*Third generation*[c]	*Fourth generation*[d]	*Other instances*[e]
Semiarid low-income countries	0.5	1.6	2.6 Mali	3.4 Somalia	Chad, Niger, Upper Volta, Mauritania, The Gambia
Other low-income countries	6.5 Malawi	31.1 Tanzania, Ethiopia, Madagascar, Mozambique	35.3 Lesotho	41.3 Zaïre, Uganda	Burundi, Rwana, Benin, Togo, Sierra Leone, Guinea-Bissau, Guinea
Oil-importing middle-income countries	17.8 Zimbabwe, Kenya, Zambia	34.8 Ghana, Côte d'Ivoire, Botswana	40.7 Swaziland	45.2 Cameroon	Mauritius, Liberia, Senegal
Oil-exporting middle-income countries		11.5 Gabon, Congo, Angola	18.4 Nigeria	21.1	

Note: The following countries are excluded: Cape Verde, Comoros, Djibouti, Equatorial Guinea, São Tome and Principe, and Seychelles, representing 4 percent of the population and 3 percent of the tractors in use, and Sudan, representing 10 percent of the tractors in use and showing a definite gradual increase.

a. Period of maximum acceleration, 1947–55; period of slowdown, 1955–81.

b. Period of maximum acceleration, 1958–70; period of slowdown, 1970–81.

c. Period of maximum acceleration, 1970–75; period of slowdown, 1975–81.

d. Period of maximum acceleration, 1975–81.

e. Number of tractors always fewer than 500 from 1947 to 1981; the data for Guinea from 1958 are completely inaccurate.

Sources: Number of tractors in service, FAO yearbooks; income levels, World Bank data.

Figure 5-3. *Tractors per Thousand Inhabitants in Use in Sub-Saharan Africa, 1947–81*

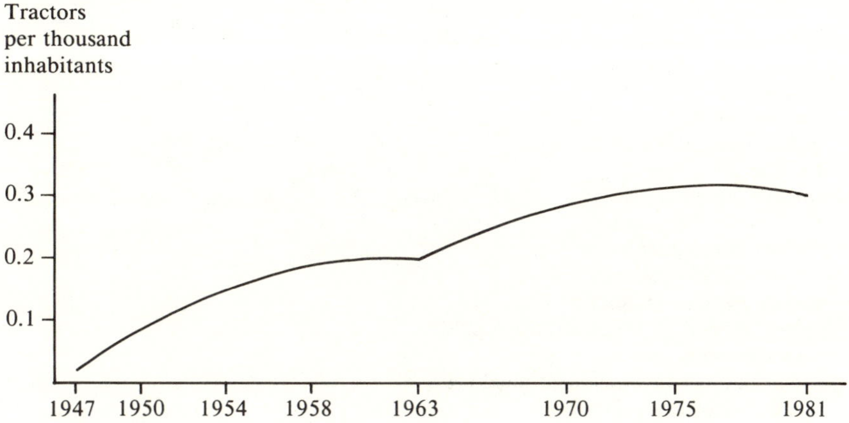

Even in the first-generation countries, the number of tractors per thousand inhabitants actually declined after reaching a peak in 1958, as shown in the following table:

1947	1950	1954	1958	1963	1970	1975	1981
0.23	0.70	1.14	1.38	1.11	1.08	1.01	0.88

The use of tractors was and continues to be restricted to a small commercial farm sector. Similar peaks in the number of tractors in use tended to occur among the second generation of tractor-using countries and in those that took up the use of tractors still later as well.

Table 5-5 summarizes the experiences of tractor projects in Sub-Saharan Africa from 1945 until the early 1980s. More details on each of these projects are presented in table 5-6. We have identified thirty tractor projects, of which seven were smallholder projects, thirteen were tractor-hire schemes, and ten were block cultivation schemes. Tractor-hire schemes are government-sponsored rental programs for multifarm use of equipment. Block cultivation schemes are groups of farms being managed and operated as a single unit, often with mechanization and other modern inputs. The following conclusions can be drawn from table 5-5: In many tractor project areas no tractors can be found today. Where any tractors are still being used, their use is inevitably associated with rice cultivation. But even these surviving tractors today are privately owned. The transition from the hand hoe to animal-draft power, where its use is appropriate, continued to be made despite the emphasis

Table 5-5. *Experiences with Tractor Projects in Sub-Saharan Africa*

Number of projects or areas	*Individual farm enterprise*	*Tractor-hire service*	*Block cultivation schemes*
Initial number of projects	7	13	10
Number of areas in which tractors are still used in the original organization form in the 1980s	None	5	None
Number of areas in which tractors are now used under private operation	None	4	2
Number of areas that had animal traction originally	3	7	2
Number of areas in which animal traction is in use in the 1980s	6	9	3
Number of areas in which continued use of the tractor is associated with rice cultivation	1[a]	9	None

a. Tractor-hire scheme.

on tractors. Of the seventeen attempts to bypass animal traction for tractorization only three succeeded, all of them associated with lowland rice cultivation. None of the block cultivation schemes has ever been successful. For an evaluation of the performance of tractors in Sub-Saharan Africa, see Labrousee (1971), Bonnefond (1967), and Kline and others (1969).

Table 5-6. *Status in the Early 1980s of Some Mechanization Attempts in Sub-Saharan Africa*

Beginning of mechanization	Country and region	Primary crop	Type of mechanization	Use of animal traction: Before 1945	Use of animal traction: Early 1980s	Status of mechanization in the early 1980s
1945	Tanzania	Groundnut	IF	+	+	TH since 1951
1947	Guinea (Siguiri)	Lowland rice	TH	+	+	Several tractors on contract
	Senegal					
1947	Boulel	Groundnut	TH	+	+	None since 1955
1947	Sedihou	Groundnut	IF	–	+	Association with animal traction in 1953, but none since 1960
1948	Fleuve	Rice	IF	–	Σ	TH
	Cameroon					
1948	Yagoua	Lowland rice	TH	–	–	Still heavily used
1948	Guetale	Groundnut	IF	–	+	No longer used
1949	Congo (Niari)	Groundnut	IF	–	–	No longer used
	Mali					
1949	Baguineda	Lowland rice	TH	+	+	Minimal (animal traction)
1949	Segou	Lowland rice	TH	+	+	Still used
1950	Sierra Leone (several)	Lowland rice	TH	–	Σ	Several tractors on contract
1950	Western Nigeria	Several	TH	n.a.	n.a.	n.a.
	Madagascar					
1951	CRAM[a]	Several	IF	+	+	None since 1951
1952	Sakay	Several	P	+	+	n.a.
1953	Northwestern Côte d'Ivoire	Lowland rice	TH	–	+	Several tractors on contract
1953	Northern Ghana	Lowland rice	TH	–	Σ	Several tractors on contract

(Table continues on the following page.)

Table 5-6 *(continued)*

Beginning of mechanization	Country and region	Primary crop	Type of mechanization	Use of animal traction: Before 1945	Use of animal traction: Early 1980s	Status of mechanization in the early 1980s
1953	Central African Rep. (Gaigne)	Cotton	P	–	–	No longer used
1953	Botswana (several)	Maize	P	+	+	n.a.
1953	Madagascar (Nanisana)	Several	TH	+	+	No longer used
1953	Zaïre (Bambera)	Cotton	P	–	–	No longer used
1956	Central African Rep. (Grimari)	Cotton	P	–	–	No longer used
1960	Kenya (Mwea)	Rice	TH	+	–	n.a.
1963	Central African Rep. (Bambari)	Cotton	P	–	–	Attempt at small-scale mechanization
1965	Cameroon (Bamenda)	Maize	P	–	+	No longer used
1965	Central African Rep. (Operation Labor)	Several	TH	–	–	No longer used
1971	Togo (Southeastern Terre de Barre)	Maize	P	–	–	No longer used
1971	Central Côte d'Ivoire	Several	P	–	–	Attempt at small-scale mechanization
1974	Guinea	Rice	TH	+	+	Still used in several areas
1977	French-speaking western Africa	Cotton	IF/P	+/–	+/–	Animal traction still used where it was introduced earlier

\+ Yes.
– No.
n.a. Not available.
IF Individual farm.
TH Tractor hire—use of tractor through leasing company.
P Paysannat—block-cultivation scheme.
Σ —use extremely limited.
a. Centre de Recherche Agronomique de Madagascar.

Part Two

Animal Draft

Chapter 6

False Assumptions concerning Animal Draft

The problems frequently cited as primary causes for the absence of animal traction in Sub-Saharan Africa fall into two general categories: a misunderstanding of the problem, such as a cultural aversion to the use of animals, or a misunderstanding of the benefits of alternative sources of power, and an overemphasis on the short-run adjustment problems, such as the development of repair services and the acquisition of animal-husbandry skills.

The perceived problems in the former category are not serious constraints, and those in the latter category can be overcome. These false assumptions have been compiled from the literature and from discussions with extension and research personnel.

Animal Traction as a Regressive Technology

A notion common among politicians in developing countries and experts, salesmen, and technicians in the industrial countries is that animal traction is an outdated and backward technology. Rapid agricultural development is often taken to imply the bypassing of the animal-traction stage and going directly from hand tools to the use of tractors and other purchased inputs such as fertilizers and pesticides. During the 1960s several newly independent African countries, among them Tanzania, Zambia, Guinea, Ghana, and Côte d'Ivoire, adopted policies that were designed to leapfrog the animal-traction stage by providing tractors and tractor-hire services at subsidized rates. Most of these attempts at rapid tractorization failed, and several countries subsequently reverted to encouragement of animal-draft power.

The tractor fiasco in Tanzania was discussed in chapter 5. A report by the Ministry of Agriculture states that there are a total of 10,000 tractors in Tanzania, of which 40 percent are unserviceable, cultivating a total area of 359,000 hectares (Tanzania 1983a). This indicates a gross underutilization of the supply of tractors. The same report also states that tractors are used on less than 6 percent of all cultivated land. Recent government policy on agricultural mechanization has begun to deemphasize the role of tractors and encourage improved hand tools and animal-drawn equipment (Tanzania 1983b).

The appropriate source of power for an agricultural operation depends on the physical and economic conditions facing the farmer rather than on the cosmetic appeal of modern machinery. In much of Sub-Saharan Africa animal traction is the necessary first step toward mechanization. We shall discuss this proposition and the empirical evidence in favor of it in chapter 10.

Lack of Mechanical Skills and Repair Services

Assertions frequently made by specialists studying agricultural mechanization in Africa are that African farmers lack the mechanical skills required for the operation of animal-traction equipment and that they do not have easy access to services for the timely repair and maintenance of equipment. We find these assertions, if not incorrect, at least misleading.

The use of bicycles, motorcycles, and mechanical mills has become common all over Sub-Saharan Africa, even in areas where animal traction is not used. The mechanical skills required for the use and maintenance of this equipment are equal to or greater than those required for animal-traction equipment. Table 6-1 shows the use of bicycles, motorcycles, and mills in areas in which animal traction is not used. In all these locations the use of one or more of these types of mechanical equipment is common despite the absence of animal-draft power.

In almost every one of the villages surveyed, whether or not animal traction is used, there is in the village or close by a blacksmith who is capable of making minor repairs on hand hoes, plows, and other animal-traction equipment. For major repairs and welding there is generally a workshop within twenty kilometers of the village. Table 6-2 shows that 80 percent of the villages surveyed have access to a workshop within twenty kilometers where major repairs can be made, and half of these workshops are less than five kilometers from the village. Table 6-2 shows further that bicycle and tractor repair facilities are also within accessible distance of the village. The problem is generally not

Table 6-1. *The Presence of Bicycles, Motorcycles, and Mills in Locations without Animal Traction*

Location	*Bicycles*	*Motorcycles*	*Mills*
Tanzania			
Kilosa district	x	?	?
Zambia			
Northern Chipata	x	—	x
Mumbwa district	x	—	?
Nigeria			
Kano state	x	x	x
Oyo state	x	x	x
Kenya			
Kikuyu highlands	—	x	x
Côte d'Ivoire			
Morondo	x	x	x

x Present.
— Not present.

the lack of repair facilities but rather the paucity of spare parts and the difficulty of obtaining them.

The problem of spare parts is greater where the equipment is manufactured outside the country or where the steel required for the plow is imported. In such instances the availability of parts fluctuates with the foreign-exchange reserves of the country, and the inventory of spare parts drops sharply during adverse economic conditions. This sequence of events has been observed in Ghana, Guinea, Tanzania, Nigeria, and Zambia. Where animal traction has been successfully incorporated into the farming system, however, it tends to survive even

Table 6-2. *Distance of Places Surveyed from Repair Facilities* (kilometers)

Item	*0–5*	*6–10*	*11–15*	*16–20*	*more than 20*	*Total*
Animal traction						
Number of instances	10	3	6	1	5	25
Percent of total	40	12	24	4	20	100
Tractors						
Number of instances	0	1	3	5	5	14
Percent of total	0	7.2	21.4	35.7	35.7	100
Bicycles						
Number of instances	7	4	4	2	7	24
Percent of total	29.2	16.7	16.7	8.2	29.2	100

under extremely adverse conditions. Approximately one fourth of our sample villages stated that they had difficulty in obtaining spare parts but they continue to use animal-drawn plows. Although lack of equipment or parts could discourage new users in the short run, it would not be expected to affect the rate of adoption of animal traction in the long run. Also, where markets function efficiently the spare-parts problem is minimized by intermediaries who reduce the transactions costs involved in obtaining equipment.

Lack of Animals and Skills in Animal Husbandry

Sub-Saharan Africa is often characterized as having a historic dichotomy between crop cultivation and livestock herding. This supposed polarization between farmers and herdsmen has been given as the reason for the slow spread of animal traction in Africa. The argument goes as follows: since the cultivators do not have any animals they lack skills in animal husbandry and are thus not in a position to maintain the animals required for traction purposes.

Although it is easy to find examples of tribes, such as the Fulani, Fulbe, Masai, Turkana, Dinka, and Pokot, that have historically been exclusively cattle herders, it is hard to find crop farmers who do not own any cattle or other livestock. The Wakara, Rundi, and Ruanda are tribes that keep sheep and goats rather than cattle, but this transition to smaller livestock has been made recently because of an extreme shortage of grazing land. Most of the tribes that lived in precolonial Africa were crop cultivators who kept cattle for milk and meat and as a store of wealth. Table 6-3 lists the principal East African tribes outside the forest zone in categories of exclusive farming, exclusive herding, or both farming and herding. Of the twenty-seven groups, twenty-one combine crop farming with ownership of cattle. The argument that farmers in Sub-Saharan Africa are not skilled in animal husbandry is for the most part invalid.

The question, then, is why some tribes who own cattle and cultivate land do not use the animals for working their land. The general answer is that these farmers face other constraints on the use of animal traction or have other opportunities for their animals. The Kikuyu in the Kenyan highlands, for example, concentrate on the production of milk and vegetables and therefore do not require the traction power that oxen can provide for land preparation. The Sukuma do not use animal-drawn plows on the hill slopes because the soil is very light and susceptible to erosion. The Meru cultivate small plots on the slopes of

Table 6-3. *Animal Husbandry Experience of Precolonial Tribes in Semiarid East Africa*

Tribe and country	*Exclusively cattle herding*[a]	*Exclusively crop farming*[b]	*Crop farming with cattle herding*	*Use of animal traction in the 1980s*
Sudan				
Suk	x		x	
Turkana	x			
Dinka	x			
Bier			x	
Kenya				
Masai	x			
Nandi			x	x
Bantu Kavirondo			x	x
Kipsigi			x	x
Kikuyu			x	
Keyu			x	
Taita			x	
Meru			x	x
Burundi				
Rundi		x		
Ruanda				
Ruanda		x		
Tanzania				
Meru			x	x
Iraqw			x	x
Sukuma			x	x
Wakara		x		
Uganda				
Bahima[c]	x		x	
Baganda			x	x
Teso			x	x
Pokot	x			
Zambia				
Tonga			x	x
Barotsi			x	x
Kamba			x	x
Lesotho				
Basuto			x	x
Botswana				
BaRolong			x	x
Tswana			x	x

a. Does not include the cultivation of garden plots.

b. Does not exclude the keeping of sheep, goats, and small ruminants.

c. This tribe is split into two groups, the pastoralists and the agriculturists.

Sources: Herskovits (1926), Ruthenberg (1980), and Schneider (1984).

Mount Meru and have a long land-preparation period, so they can complete all their work without the use of animal power. Where the conversion to animal-draft power has been profitable, several tribes, such as the Machakos, the Tonga, the Tswana, and the Teso, have made it.

Can farmers who never owned cattle, such as the ones in the tsetse belt, acquire skills in animal husbandry and use animal-draft power? Presumably such farmers have greater difficulty in acquiring these skills, but there are many examples of their having done so. Consider the farmers in the Mumbwa district of Zambia. Until recently this was a sparsely populated, tsetse-fly–infested area. As the density of population increased the forest cover receded, and large tracts of land became free of the tsetse fly. Consequently, the cattle population rose and the use of animal traction is emerging rapidly in that area today. Similar examples can be found in Sierra Leone, Southern Nigeria, Côte d'Ivoire, and Guinea. The historic absence of cattle has not prevented the spread of animal traction in these areas when conditions were appropriate. Where animal-traction projects failed to inculcate such skills among the target population it was not because the population could not learn but because the costs of acquiring these skills were greater than the benefits of switching to animal power, given the low intensity of the existing system of farming.

Cultural Differences

Another explanation commonly offered for the lack of animal traction in a given area is that the people have a cultural aversion to the use of animals for work. While it is true that in areas where cattle herding or livestock keeping has not been prevalent the adoption of animal traction is slower because of the costs of learning, we have been unable to find any cultural barriers to the use of animal-draft power.

On the other hand, several examples of the use by the same tribe of animal traction in one area but not in another were found. The Sukuma, for example, use animal-drawn plows in the valleys while at the same time cultivating the hill slopes with hand-made ridges. Similarly, the Meru of Mount Kenya use animal traction only in the valleys. Tribes that do not use animal traction, such as the Wakara, the Iraqw, and the Kikuyu, generally switch to draft power use when they migrate to areas where its use is prevalent. Several pastoral tribes, such as the Fulani, the Fulbe, and the Masai, have in recent years adopted crop cultivation and use animal power despite a historic aversion to such a sedentary way of living.

In order to show conclusively the effect of cultural aversion it would be necessary to find an area in which animal traction is prevalent, yet some tribal group, under the same physical and economic conditions, continues to cultivate exclusively by hand.

Significance of Oxen Size

Trypano-tolerant cattle in Sub-Saharan Africa are generally smaller than the Zebu cattle and are therefore thought less suitable for work. These cattle have somewhat lower tractive power and lower endurance than Zebus. The N'dama and the Baoule are the smallest breeds and are therefore likely to have the lowest tractive power. It is, of course, clear that, other things being equal, this fact will make animal traction less attractive or more costly. Nevertheless, this has not prevented these breeds from being used for traction. Where greater traction is required several pairs of oxen are teamed to obtain the necessary power. There is no scientific proof of the assertion that endurance of cattle is proportional to size. According to Munzinger (1982), endurance depends on the constitution and condition of the animal, the guidance and treatment of the animals by the person driving the team, the nature of the draft work, and the speed. Endurance therefore appears to be governed by the environment rather than by the breed (Munzinger 1982, 77). A well-maintained team of N'dama or Baoule oxen would probably have endurance equal to that of the average Zebu.

Crossbreeds between these smaller cattle and the Zebus, such as the Borgou and Mere, are larger and have greater tractive power and could therefore be used as an alternative to a team of several pairs of small oxen. Breeding activities aimed at producing suitable beef and dairy cattle have also produced fairly heavy cattle that are suitable for draftwork. Such breeding experiments in Kenya have been successful, and improved breeds of oxen are now used in Kakamega, Bungoma, and other districts of western Kenya (Jaetzold and Schmidt 1982).

Lack of Fodder

The availability of fodder has not usually been a constraint to the adoption of animal traction. This is because of the positive relation between intensification of the agricultural system and the availability of fodder, at least in the early stages. During the forest-fallow stage there are almost no grazing areas, but as reduction in fallow periods reduces the forest cover the amount of grassy land increases and a consequent increase in ownership of cattle is observed. Around the bush-fallow and grass-fallow stages of cultivation abundant communal grazing areas are available with few restrictions by the members of the community on the rights of users. Further intensification of the agricultural system to the permanent-cultivation stage, however, leads to competition between using the land for grazing and using it for crops. It is at this stage that restrictions on communal grazing rights and the emergence of private grazing areas begin to be observed. As the scarcity of land increases crop residues are substituted for grass as fodder, and eventually fodder crops are introduced into the crop-rotation cycle.

Competition for land and the ensuing shortage of grazing areas are found at very high intensities of cultivation. Table 6-4 presents a two-way relation between farming intensity and fodder-supply systems derived from data gathered during field visits to Africa and India. Communal grazing is the predominant system in thirty of the forty-seven locations in Sub-Saharan Africa. The highest number of instances of communal grazing is found at the short-fallow stage; from then on private grazing begins to emerge. All but one of the instances of private grazing are in the annual-cultivation stage, the one exception belonging to an area that is in transition to annual cultivation. We found no instances in Sub-Saharan Africa of livestock being maintained exclusively on crop residues and fodder crops. By contrast, in four of the five instances from the semiarid tropics of India, intensive fodder systems were in use; fodder crops were being grown and all crop residues were being collected from the fields. The fodder supply in India is extremely limited and there is a well-defined market for fodder, yet the use of animal draft power in Indian farming systems is widespread. None of the farming systems of Africa has reached an intensity equal to that in India, so fodder management is not yet a critical problem and does not limit the number of oxen kept for draft purposes.

Table 6-4. *Changes in Fodder-Supply Systems with Changes in Farming Intensity*

Farming system	*Communal grazing*	*Communal grazing and supplementary feeding*	*Private grazing*	*Private grazing and supplementary feeding*	*Fodder production and crop residues*	*Not applicable*	*Total number of instances*
Forest-fallow	2	1				2	5
Bush-fallow	4	2				2	8
Short-fallow	4	9					13
Short-fallow with emerging annual cultivation		6	1				7
Annual cultivation (Africa)	2	5	1	5		1	14
Annual cultivation (India)				1	4	5	
Total number of instances	12	23	2	6	4	5	52

Chapter 7

Farm-Level Benefits of the Transition to the Animal-Drawn Plow

The most commonly cited benefits of the switch to the animal-drawn plow are expansion of the area cultivated, greater yields, reduced labor requirements, and the development of animal-husbandry skills. There are also possible negative effects, such as an increase in socioeconomic inequality. In our review of the empirical literature we identified twenty-two surveys in which an attempt was made to evaluate the effects of animal traction. These studies, which were contemporaneous comparisons of holdings in the same agroclimatic environment with and without the use of animal traction, are summarized in table 7-6 below. We are not aware of any long-term comparisons of households that have switched from hand hoes to animal draft power.

The primary objection to pure cross-sectional comparison between farms on which animal traction is used and those still cultivated with the hand hoe is that they differ in many respects besides the sources of power that they use. Animal-traction farms are usually larger and wealthier and have better access to credit markets. These differences affect both the decision to use the plow and the decision to use other advanced techniques that are unrelated to animal traction. Households that use animal traction would therefore be expected to be more productive even without access to animal traction, and the advantages of the plow would be exaggerated in a pure cross-section comparison. This problem is discussed in the appendix to this chapter, where some guidance is offered to future researchers about ways of overcoming it. The empirical studies considered here are the ones in which substantial care was exercised to separate the effects of the use of animal traction from other effects, such as those from the use of fertilizers and high-yielding varieties, changes in agronomic practices, and so on. Even in these instances, however, not all other things are equal.

Area Effects

In comparisons of hand-hoe and animal-traction farms the latter were generally found to be larger than the former. In all seventeen of the studies that provide data on area effects, it was reported that farms on which animal traction was used were larger than those on which it was not (see table 7-1). The average size of holding on which animal traction is used is 6.6 hectares, while that of a holding cultivated manually is only half as great—3.3 hectares. When farming conditions were classified as favorable—more than 750 millimeters of rainfall a year—and unfavorable—less than 750 millimeters of rainfall a year—animal-traction farms were still found to be twice as large as hand-hoe farms. Similarly, animal-traction farms have a greater area per person than those on which animal traction is not used; the former have 25 percent more land than the latter, irrespective of the zone they are in. These findings suggest that the transition to the plow is associated with an expansion of the private area cultivated and therefore an increase in agricultural production with the use of animal traction. Whether there are also an expansion of the aggregate area and a production effect depends on additional factors.

The aggregate area cultivated will expand only when a substantial portion of the land that animal-traction households expand into is fallow or reclaimed land. If, on the other hand, expansion of animal-traction households takes place at the expense of other farmers then little or no increase would be observed in the aggregate cultivated area. The following examples illustrate this point.

Consider farmer A, who replaces his hand hoe with a plow on land that he cultivated himself the preceding year. As long as he does not improve his tillage, his yield will not be greater; by definition, he does not have an area-expansion effect. The main benefit to him of a switch to the plow is a reduction in the labor used for land preparation. Now suppose that A expands his area by renting from farmer B land that was cultivated with a hoe the preceding year. The result is a private area effect but no aggregate area effect—that is, the total cultivated area does not increase; land is merely transferred from one cultivator to another.

Now suppose that A, instead of renting from B, expands into fallow land or uses the plow on land never cultivated before. In addition to the yield effect, A realizes an absolute area-expansion effect; both A's private area and the aggregate area under cultivation in the region increases. Finally, suppose that A increases the rate at which he uses his plow by cultivating a second crop each year on the same plot of land. The result is both private and aggregate area-expansion effects.

Table 7-1. *Summary of the Evidence of the Effects of the Use of Animal Traction*
(number of examples)

Variable	*Examples measured (out of 22)*	*Positive effect*	*No significant difference*	*Negative effect*	*Area effect*
Area per farm					
All examples	17	17			6.6 hectares with animal traction, 3.3 without
Examples in favorable areas	9	9			6.8 hectares with animal traction, 3.5 without
Examples in unfavorable areas	8	8			6.3 hectares with animal traction, 3.0 without
Area per person					
All examples	19	19			
Examples in favorable areas	11	11			0.68 hectares with animal traction, 0.55 without
Examples in unfavorable areas	8	8			0.58 hectares with animal traction, 0.45 without
Use of land for market crops					
All examples	19	12	7		
Examples in favorable areas	11	8	3		
Examples in unfavorable areas	8	4	4		
Yield per hectare					
All examples	14	4	8	2	
Examples in favorable areas	10	4	6		
Examples in unfavorable areas	4		2	2	

Note: Favorable areas are those in which annual rainfall exceeds 750 millimeters; unfavorable areas are those in which annual rainfall is less than 750 millimeters and upland farming is practiced.

It can thus be concluded that the transition from the hand hoe to an animal-drawn plow leads to an expansion in area cultivated only when private farmers extend cultivation into fallow land, reclaim previously uncultivated land, or increase their farming intensity from annual cultivation to multiple cropping. An intuitively appealing hypothesis would be that in land-scarce economies, private area expansion is primarily at the expense of other farmers, while in land-abundant economies it is primarily at the expense of fallow or uncultivated land.

Empirical evidence is widely available on differences in the size of farms with and without animal traction. The subjective evaluations we conducted were in agreement, all thirty showing a positive area effect. We are not aware of any studies conducted in Africa, however, in which an examination of the expansion in area was made to determine whether it was at the expense of other farmers or of virgin, fallow, or grazing land. Recall data or before-and-after studies would be needed if such a determination were to be made. From the available evidence, therefore, no inference on private versus aggregate area effects can be drawn apart from our hypothesis. In Africa aggregate area-expansion effects are more likely to be observed. John McInerney and Graham Donaldson's is the only study in which this issue has been examined. They concluded that in Pakistan the expansion in area by tractor-using farmers came mainly through the transfer of control of land already under cultivation. They found that 78 percent of all expansion came from purchase, rental, or displacement of tenants, while only 22 percent came from reclamation of land (McInerney and Donaldson 1975, 37).

A question related to that of where the additional area comes from is, What crops are grown on the additional area? In other words, is animal-draft power associated with a replacement of the production of subsistence crops with production of market crops? Of the nineteen instances in which an expansion in cultivated area per person was observed, in twelve the additional area was being used to produce market crops, such as cotton, groundnut, and rice. More important, on animal-draft farms as great an area was planted in millet or sorghum, the traditional subsistence crops in Africa, as on hand-hoe farms. A possible inference is that where the additional area comes from fallow land or reclaimed land the spread of animal-drawn plows leads to an expansion in the production of commercial crops. The expansion in area on which commercial crops are grown is more likely to occur in the favorable zones than in the unfavorable zones.

Yield Effects

We begin by noting that—apart from subsoiling—it is almost always technically feasible to achieve a given quality of tillage by using either a plow or manual techniques. Whether achieving a given quality of tillage by hand is economically justifiable is of course another question.

Differences of treatment in land-preparation experiments are differences in the tool used to prepare the land—the hand hoe or the plow, or one plow instead of another. Tillage quality varies according to which of these tools is used. A difference between the result of using, say, the hand hoe and using the plow is therefore called the effect of the plow. But since it is technically feasible to achieve essentially the same tillage quality by hand, the treatment effect is better considered a tillage-quality effect.[1]

Experiment station results show that the effects of improved tillage vary with soil type and crop. The yield response to improvement in tillage, for instance, is greatest on heavy soils and least on light sandy soils (see chapter 4 for details).[2] On a given type of soil the yield benefits of tillage are greatest for rice and least for groundnuts. For recent surveys of experiment station tillage results, see Bloom (1979) and Sargent and others (1981). Despite these experimental results, survey research to compare yields achieved on hand-hoe farms with those achieved on animal-traction farms have consistently shown a minimal yield effect.

Yield increases associated with a switch from the hand hoe to the animal-drawn plow can, of course, occur only when the switch is followed by improvements in the quality of tillage. The following examples, summarized in table 7-2, illustrate some of the ways in which quality of tillage can be increased:

- A plot of land that was directly seeded before is now tilled before seeding. In this instance a yield response to tillage would be expected regardless of whether the tillage was done by hand, animal-drawn plow, or tractor-drawn plow.
- On a plot of land on which deep plowing produces yield effects, shallow plowing is replaced with deep plowing.
- A change of plow—from a scratch plow to a moldboard plow, for example—could improve yields because the soil is not turned over, crop residues are incorporated better, and weed control is improved.
- A more timely completion of land preparation leads to better yields.

In theory, it is possible to identify quality differences between animal-draft and hand-hoe farms by means of cross-section studies.

Table 7-2. *Expected Effects of a Shift from the Hand Hoe to the Animal-drawn Plow or the Tractor*

Variable	*Potential yield effect*	*Labor use for land preparation*	*Private area effect*	*Aggregate area effect*
Improvement in quality of land preparation on a specific plot				
From direct seeding to tillage with a hand hoe	+	+	0	0
From direct seeding to the plow	+	+	0	0
From one type of plow to another	+	?	0	0
More timely completion of land preparation	+	0	0	0
Constant quality of land preparation with a shift in source of power				
On land previously cultivated by the same farmer	0	−	0	0
On land formerly cultivated by others	0	−	+	0
On land formerly fallow or uncultivated	+	+	+	+
Multiple cropping on the same plot	+	+	+	+

\+ Positive effect.
− Negative effect.
0 No significant difference.
? Effect unknown.

None of the studies, however, will tell us what quality of tillage prevails on hand-hoe farms. We do not know, for example, whether any direct seeding takes place. Many of the studies also do not tell us what quality of tillage occurs on animal-traction farms. Plots still prepared by hand, for example, may be included in determining average yields. From the existing studies we cannot really tell whether the shift in source of power is associated with any improvements in quality. We can only make inferences from the available evidence.

The yield effects of the animal-drawn plow were examined in fourteen of the twenty-two examples presented here (see table 7-1). Of these, eight show no significant differences in yield, two show a negative effect, and four show a positive effect. That there is no difference in yield in the majority of examples leads us to infer that there is no difference in the quality of tillage between that on hand-hoe and that on animal-draft farms. In other words, in most instances farmers who switch to animal traction do not improve the quality of tillage in any of the four ways mentioned earlier. This is in accordance with the world-wide finding that the switch from animal traction to tractors has little or no effect on yields (Binswanger 1984 and 1978; Herdt 1983). While in many instances it may be possible to increase the quality of tillage by using a different source of power, farmers rarely seem motivated by such potential gains in yield at the time of the switch. The switch in the source of power, therefore, must be motivated by other considerations, such as expansion of the area cultivated or a saving in the amount of labor required.

Let us now examine the examples in which significantly positive differences in yield were reported and find out what was so different about them. Consider first the study made in Western Kenya by Oluoch-Kosura (1983); he found significantly greater maize yields for farmers who used animal-drawn plows. He also found that farmers who owned both oxen and plows increased their maize acreage by expanding into fallow land. Since fallow land is more fertile, it may provide larger yields, so the average yields per hectare of farmers who use animal draft are larger than those of farmers who use the hand hoe.

Consider next the case of the eastern regional development organization in Burkina Faso, studied by Michigan State University. Farmers using animal traction are reported to have achieved significantly larger maize yields than those using the hand hoe (Barrett and others 1982). The causality between tillage with a plow and larger maize yields has not been established, since four times as much per hectare is spent for chemical fertilizers on animal-traction farms than on hand-hoe farms (Lassiter 1981, 32). The higher yields attributed to animal traction may in fact be attributable to the use of fertilizers.

Finally, consider the two instances reported by Jaeger (1984)—Diapangou in eastern Burkina Faso and Boromo in central Burkina Faso. Animal-traction farms in both areas are reported to have significantly larger yields of white sorghum. These instances are influenced by the same confounding effects of fertilizer use mentioned above. In Boromo ox-traction farmers use significantly higher levels of organic and chemical fertilizers than do hand-hoe farmers. Similarly, the amount of fertilizer used per hectare on animal-traction farms is eight times as large as that used on hand hoe farms in Diapangou (Lassiter 1981, 27).

Let us now consider the two instances in which a significant decline in yields on animal-traction farms was reported. Both these are in central Burkina Faso, on the Mossi plateau, and on both donkey traction is used (McIntire 1981b, Jaeger 1984). Unlike ox traction, donkey traction in Burkina Faso is used mainly for weeding. In Nedogo, central Burkina Faso, for example, 75 percent of the total working time of donkeys—239 hours—is used on weeding and only 6 percent on primary tillage. In Diapangou, however, 61 percent of the total working time of oxen—136 hours—is spent on primary tillage (Jaeger 1986). A yield response to tillage, therefore, cannot be expected in this instance. The reason for lower yields per hectare could be the larger area cultivated and therefore less intensive care per hectare on donkey-traction farms.

Even if all other tillage-quality factors remained the same on both hand-hoe and animal-draft farms the timeliness effect could still be expected to hold. A switch to the plow ought to reduce the time required for land preparation and thus make it possible to avoid delays in seeding.[3] Why do we not therefore observe substantial gains in yields? There are two possible reasons. First, the length of the sowing period is directly related to the amount of annual rainfall; timeliness is therefore a crucial factor only for rainfed cultivation in the semiarid zones in which rainfall is low—locations we have categorized as unfavorable. In the favorable zones there seems to be a sowing period of at least a month or six weeks during which yields do not decline substantially. Second, if animal traction leads to area expansion, and if the limit to this expansion is the duration of the sowing period, the average timeliness of seeding may not differ on hand-hoe farms from that on animal-traction farms. Unfortunately, in no study in Sub-Saharan Africa has the distribution in time of sowing dates of animal-traction and hand-hoe farms actually been measured. The only existing study of this is that of Kahlon (1975), who examined the timeliness of sowing on tractor farms in relation to that on animal-traction farms in the Indian Punjab. He found no differences between the two groups in the distribution of sowing dates.

Table 7-3 summarizes the subjective evaluations of the effects of animal traction. When farmers were asked directly the very simple question whether they expected larger, smaller, or the same yields from plots tilled by hand hoes as from those tilled by plows, the most common response—twenty-two out of thirty—was that they expected an increase in yield. This may not be surprising, since the question was not sufficiently precise to force the farmer to hold all the confounding factors constant in his mind when making the comparison.

Effects on the Use and Productivity of Labor

There is general agreement that a transition from the hand hoe to the plow reduces the amount of labor required during the land-preparation period.[4] Ruthenberg (1980) reviews studies in which the use of labor, by operation, on hand-hoe farms is compared with that on animal-draft farms.

Table 7-3. *Subjective Evaluations of the Effects of the Use of Animal Traction*
(number of interviews)

Variable	*Increase*	*No change*	*Decrease*
Area	30	0	0
Yield	22	8	0
Labor requirement per hectare			
For land preparation	5	9	16
For seeding	7	22	1
For weeding	12	8	10
For harvesting	14	14	2
Total labor	10	12	8

Note: Of forty-six interviews conducted, thirty were applicable to this survey. Of these, two were from Botswana, two from Tanzania, four from Zambia, five from Kenya, six from Cameroon, three from Nigeria, two from Côte d'Ivoire, three from Guinea, and three from Burkina Faso.

There is widespread disagreement, however, as to whether the total labor requirement per hectare declines when the plow is used. First let us consider the subjective evaluations. More than half of the thirty villages report a decrease in the labor required for land preparation when the hoe is replaced with the plow. Only eight of the thirty, however, report an overall decrease in labor requirements. Recent studies have suggested that a shift in the labor peak from land preparation to weeding or harvesting may cause the overall labor requirements to rise rather than fall (Norman and others 1982, Bigot 1981, Delgado 1979b, Kline and others 1969).

While it is true that expansion of the area under cultivation leads to greater requirements for weeding and harvesting when these operations continue to be performed by hand, it is also clear from the survey of research studies that the labor required per hectare declines when the hand hoe is replaced by the animal-drawn plow. McIntire (1984), reviewing nine surveys of the use of animal traction in West Africa, summarizes the situation as follows: "The general outcome is to raise labor per farm and to reduce it per hectare."

Table 7-4 presents the results of an analysis of the patterns of the use and productivity of labor with agricultural intensification and mechanization. A worldwide data set, consisting of hand-hoe, animal-draft, and tractor farms from fifty-two specific locations in Africa, Asia, and Latin America, was assembled for this analysis from a variety of published sources (Pingali and Binswanger 1984). Labor input was measured in hours per hectare and yield was measured both per hectare

Table 7-4. *Use and Productivity of Labor with Intensification and Mechanization of Agriculture*
(logarithm)

Variable	*Log labor use*	*Log yield per hectare*	*Log yield per hour of cultivation labor*
Log farming intensity[a]	0.456*	−0.389*	−0.068
	(0.132)	(0.097)	(0.15)
Animal traction[b]	−0.96*	−0.179	0.78*
	(0.23)	(0.17)	(0.26)
Land investment	0.057	0.299*	0.24
	(0.20)	(0.148)	(0.23)
Intercept	5.46	5.69	0.24
	0.33	0.45	0.38
Number of observations	56	56	56

*Significant at 1 percent.
a. Farming intensity is the frequency with which a plot of land is cultivated.
b. Intercept shifters.
Source: Pingali and Binswanger (1984).

and per labor-hour. The independent variables were farming intensity, defined as the frequency with which a plot of land is cultivated, and dummy variables of animal traction, tractor use, and land investment, which includes destumping, leveling, drainage, and irrigation.[5]

A log-linear regression of labor use showed a significant positive association of farming intensity with labor used per hectare. Also, a switch from the hand hoe to animal-draft power and to the tractor significantly reduced the total labor use per hectare. In other words, the transition to more intensive farming systems, holding technology constant, is associated with greater labor input per hectare. The graduation from the hoe to other mechanical technologies leads to substantial savings in labor use (see figure 7-1).

Regressions were also performed on yield per hectare and per labor-hour. We found that while yield per hectare rose with farming intensity, there was no significant association of yield with the use of animal traction or the tractor. A change in mechanical technology, however, had a significant positive effect on yield per hour of cultivation labor, which does not include the overhead labor required for animal maintenance or land investments. We can conclude, therefore, that although the animal-drawn plow and the tractor have little or no effect on yield per hectare, they have a substantial beneficial effect on the productivity of labor, mainly through a reduction in the amount of labor required for

Figure 7-1. *Labor Use and Farming Intensity, Logarithmic Scale*

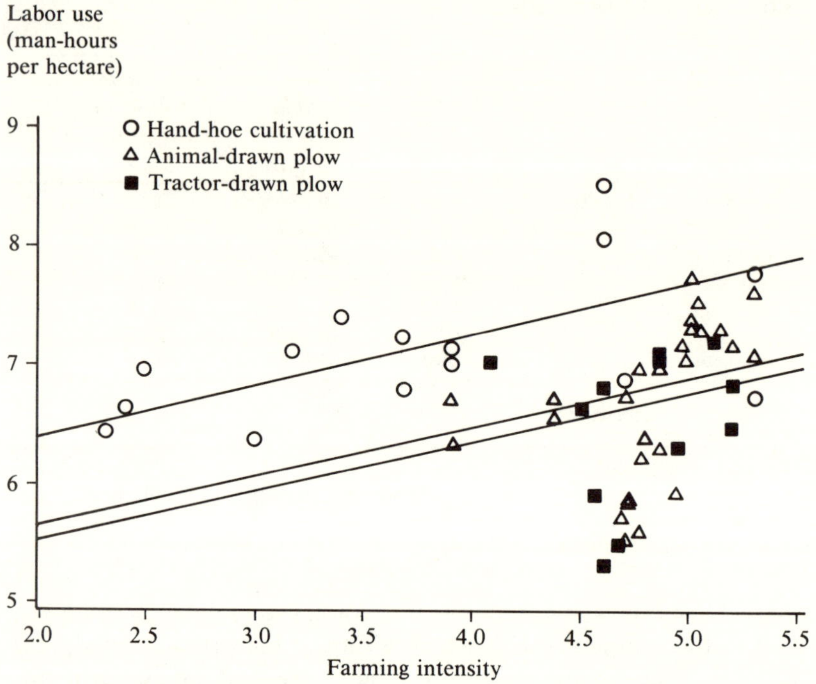

cultivation. Since our data set did not include the overhead labor required for the training and maintenance of draft animals and for the land investments, such as destumping and leveling, associated with intensification, we are not in a position to decide whether intensification, including the associated shift in tools, leads to an increase or a decrease in the overall productivity of labor where the overall labor inputs include overhead labor.

Effects on Livestock Raising

Is the use of animal-draft power complementary to, or does it conflict with, livestock-raising activities? To answer this question adequately would necessitate consideration of the following three distinct situations: smallholder mixed farming, smallholder crop farming with livestock entrustment, and crop farming in areas in which livestock do not exist.

Smallholder Mixed Farming

The extent of integration of livestock into crop farming varies with agricultural intensification. In sparsely populated areas livestock raising is usually separated from crop production. As the agricultural system intensifies livestock become an integral part of crop farming, for the maintenance of soil fertility with the use of manure and for labor saving through animal-draft power. In addition to its use for plowing, animal-draft power is used for transporting manure to the fields and final products to the market, and the importance of crop residues as a source of fodder increases. The use of animals for draft is certainly a complementary activity to livestock raising in a mixed farming environment. The farmer is already familiar with animal husbandry, crop residues are available as supplementary feed, the labor required where animal-draft power is not used is substantially greater for the same level of intensity, and young oxen are easily available for training. The slaughter value of an animal rises, moreover, when it has been used a few years for work. The high-rainfall, semiarid zone and the subhumid zone are ideal for such integration of crops and livestock (Jahnke 1982, 129).

Crop Farming with Livestock Entrustment

In the low-rainfall, semiarid zones of West Africa known as the Sahel, farmers typically entrust their cattle to herding groups of the Fulani tribe who care for them the year round. The high seasonal labor requirements for grazing and supervising cattle and the high degree of skill required for transhumance provide an economic explanation of the reason farmers entrust their animals to herders rather than tending them themselves (Delgado 1979b; Delgado and McIntire 1982). Under such circumstances the adoption of animal-draft power increases the amount of work required of the crop-farming household, especially during the slack season. This in itself, however, is not a sufficient basis for the conclusion that animal traction conflicts with other livestock-raising activities in these communities. If the returns on animal-powered plowing or transport are high, then a farmer may find it profitable to maintain a pair of animals the year round—despite the high cost of searching for fodder or water in the dry season—while the rest of his stock is still herded by specialized herders. Indeed, Delgado and McIntire (1982) do find that returns on the production of cash crops—peanut and cotton—using ox-drawn plows more than offset the costs of the labor for animal maintenance. Where the use of animal-draft power is profitable in the Sudano-Sahelian farming systems, it complements rather than conflicts with livestock raising because crop farmers acquire livestock-husbandry skills and the area-expansion benefits of the use of animal traction lead to net gains in income for crop farmers.

Areas in Which Livestock Have Been Absent

In Sub-Saharan Africa, cattle and other livestock have historically been absent in areas of tsetse fly infestation (see chapter 2). The introduction of animal-draft power in these areas, under appropriate intensities, leads to the acquisition of animal-husbandry skills in addition to the other benefits of animal power. Trypano-tolerant cattle, such as the N'dama, have been successfully introduced in these areas for draft and for meat production.

The Income Effect

Does the adoption of animal-draft power lead to higher net incomes than are realized on hand-hoe farms? The income effect is the combination of area, yield, and labor-savings effects. It is generally measured in one of the following ways: net return per household, net return per hectare, and net return per worker.

In six of twenty-two examples net returns per household were reported and households using animal-traction were found to have higher net returns than those using the hand hoe in all six instances. In seven of twenty-two examples the net returns per hectare were examined and animal-traction households were found to have significantly higher returns in all but one instance. In eleven of twenty-two examples the net returns per worker were examined, and these were found to be significantly higher for animal-traction farms in six instances and not significantly different in five. Table 7-5 provides several examples from Mali and Burkina Faso of the net revenue with and without the use of animal traction.

The common element in all these examples is a significantly larger net revenue per household; animal-traction farms have between two and five times the net revenue of hand-hoe farms. Positive differences are recorded for revenue per hectare and per capita, but because households that use animal traction typically have more members and farm larger areas, these differences are not nearly so large as those in net revenue per household.

Table 7-5. *Examples of Returns on the Use of Animal-Draft Power (CFA francs)*

		Households not using animal traction			*Households using animal traction*		
Location	*Year*	*Net revenue per household*	*Net revenue per hectare*	*Net revenue per worker*	*Net revenue per household*	*Net revenue per hectare*	*Net revenue per worker*
Mali-Sud	1978	216,794	74,756	54,267	1,406,252	95,859	84,984
Eastern Burkina[a]	1979	87,315	22,409	n.a.	144,716	20,296	n.a.
Central Burkina	1980	n.a.	37,834	n.a.	n.a.	56,717	n.a.
East Burkina	1982	103,804	24,234	22,086	172,953	20,604	26,205
Central Burkina	1982	88,123	22,153	20,028	206,056	26,328	29,022
Central Burkina	1982	149,322	38,961	28,174	503,011	43,510	43,363
North Burkina	1982	64,530	13,399	14,666	128,493	15,121	14,277

n.a. Not available.

a. Burkina Faso.

Sources: For Mali-Sud, Whitney (1981); for east Burkina 1979, Barrett and others (1982); for central Burkina 1980, McIntire (1981b); for east, central, and north Burkina 1982, Jaeger (1986).

The Concentration of Income and Wealth

As mentioned earlier, animal-traction farm households typically have more members, farm larger areas, and have greater wealth. Clearly the per capita differences in area farmed and in wealth are smaller than the differences per household. Nevertheless, the question remains whether animal traction contributes to greater inequality. The survey studies do not provide conclusive answers to this question precisely because they did not follow the farms throughout a long period. But a few comments are in order.

Larger household size and greater wealth are usually associated with a higher degree of skill in management on the part of the head of the household. These skills, and the associated wealth and differences in the labor force are part of the reason for the adoption of animal traction. To some extent, therefore, the observed cross-sectional differences across households are not the consequence, but the cause of the use of animal traction.

Nevertheless, where its use is profitable, animal traction will give its early adopters a competitive advantage over nonadopters. If output expands sharply, prices will eventually drop, and the early adopters will have reaped virtually all the innovators' rents. The early competitive advantage may also allow the animal-traction farms to expand or secure their ownership of previously fallow or uncultivated land. They will typically emerge as demanders of labor in the labor market. Having already been identified as larger households, they will probably have been given preferential treatment in extension and credit services even before they began using animal traction. But area expansion will further that advantage. It is therefore clear that social and economic differentiation should be expected to increase with the introduction of animal traction. Whether that increase is permanent or temporary depends on whether the late adopters are eventually able to take full advantage of animal traction themselves. Given the size of the investment required, this is considerably less likely than would be true of a highly divisible innovation such as seeds.

Two further points need to be made. If animal traction is profitable and leads to considerable agricultural growth, nonfarm opportunities in the area will improve as a result of it, providing better employment opportunities for the less advantaged. Second, the alternative to animal traction may not be continued hand-hoe cultivation, but may be either agricultural stagnation, which benefits no one, or the introduction, at a later stage, of tractors. Because tractors are much lumpier investments than draft animals, their introduction is likely to generate far greater social differentiation than animal traction.

Table 7-6. *Effects of the Introduction of Animal Traction, Selected Countries and Regions, Selected Years, 1959–83*

Year	Country	Region	Institution	Draft animal generally used	Main tool	Average age (years)	Main crop	Total farm area (hectares): With animal traction	Total farm area (hectares): Without animal traction
1959	Mali	Central delta	INSEE	Ox	Plow	10	Rice	13.4	5.4
1960	Senegal	Groundnut area	INSEE	Horse	Seeder	5	Groundnut	6.2	2.8
1960	Mali	Haute valley	INSEE	Ox	Plow	5	Sorghum	11.5	4.5
1962	Senegal	Laghem	BDPA	Horse	Seeder	7	Groundnut	10.6	4.5
1960–66	Chad	South	INSEE	Ox	Plow	5	Sorghum	4.8	3.0
1964	Cameroon	North	CFDT	Ox	Plow	5	Sorghum	5.25	3.0
1964	Burkina Faso	Yatenga	BDPA	Ox	Plow	5	Millet	2.8	1.0
1966[a]	The Gambia	Various	Wye College	Ox	Furrow	5	Groundnut	n.a.	n.a.
1973	The Gambia	Various	ODM	Ox	Furrow	10	Groundnut	n.a.	n.a.
1974[b]	The Gambia	Various	Mettrick	Ox	Furrow	10	Groundnut	10.2	6.8
1974	Cameroon	North	Fournier	Ox	Plow	15	Sorghum	n.a.	n.a.
1976–78	Côte d'Ivoire	North	IDESSA CFDT	Ox	Plow	3	Maize	9.0	5.0
1978[c]	Mali	South	Purdue	Ox	Plow	10	Rice	5.2	2.6
1979[d]	Burkina	East	Michigan	Ox	Plow	2	Millet	3.3	2.0
1978–82	Côte d'Ivoire	North	IDESSA	Ox	Plow	5	Cotton	6.0	3.0
1980[e]	Burkina Faso	Central	ICRISAT	Donkey	Hoe	15	Millet	3.5	5.6
1981[f]	Kenya	West	Oluoch	Ox	Plow	40	Maize	n.a.	n.a.
1982[g]	Burkina	East	FSU	Ox	Plow	5	Millet	4.0	2.0
1982[g]	Burkina	Central	FSU	Donkey	Hoe	15	Sorghum	4.3	2.2
1982[g]	Burkina	Central	ICRISAT	Ox	Plow	5	Sorghum	5.2	2.1
1982[g]	Burkina	North	ICRISAT	Ox	Plow	5	Millet	4.5	2.4
1983[h]	Sierra Leone	Northwest	Oxen project	Ox	Plow	50	Rice	n.a.	n.a.

(Table continues on the following page.)

Table 7-6 *(continued)*

Area per person (hectares)		Land use				Area worked		
With animal traction	*Without animal traction*	*Crop*[i]	*Area*[j] *(percent)*	*Yield*	*Crop*	*By team (hectares)*	*On contract (percent)*	*Agro-climatic suitability for crops*[k]
1.2	1.0	Rice	+23	NSD	Rice	n.a.	n.a.	Favorable
0.60	0.45	Groundnut	+45	n.a.	n.a.	n.a.	n.a.	Unfavorable
0.70	0.40	Various	+64	n.a.	n.a.	n.a.	n.a.	Unfavorable
0.90	0.75	Groundnut	+20	NSD	Groundnut	n.a.	n.a.	Unfavorable
0.60	0.55	Cotton	+8	n.a.	n.a.	4.55	45	Unfavorable
0.75	0.60	Cotton	+10	n.a.	n.a.	4.15	43	Favorable
0.35	0.25	NSD	NSD	NSD	NSD	5.0	33	Unfavorable
0.75	0.60	Groundnut	+22	NSD	Groundnut	n.a.	n.a.	Favorable
0.50	0.40	Groundnut	+15	n.a.	n.a.	n.a.	n.a.	Favorable
0.75	0.65	Groundnut	+14	n.a.	n.a.	n.a.	n.a.	Favorable
n.a.	n.a.	n.a.	n.a.	n.a.	n.a.	5.55	50	Favorable
0.65	0.55	Cotton	+43	NSD	Maize, cotton	n.a.	n.a.	Favorable
0.55	0.35	Rice	+55	NSD	Rice	n.a.	5	Favorable
0.70	0.60	NSD	NSD	SD+	Maize	n.a.	n.a.	Favorable
0.60	0.45	Cotton	+100	NSD	Maize, cotton	5.0	n.a.	Favorable
0.35	0.30	NSD	NSD	SD–	Millet	n.a.	n.a.	Unfavorable
n.a.	n.a.	n.a.	n.a.	SD+	Maize	n.a.	n.a.	Favorable
0.55	0.45	Millet		SD+	Millet	n.a.	n.a.	Favorable
0.65	0.45	NSD	NSD	SD–	Sorghum	n.a.	n.a.	Unfavorable
0.45	0.40	NSD	NSD	SD+	Sorghum	n.a.	n.a.	Favorable
0.50	0.55	Millet	n.a.	NSD	n.a.	n.a.	n.a.	Unfavorable
n.a.	n.a.	n.a.	n.a.	NSD	Rice	6.0	37	Favorable

n.a. Not available.
NSD No significant difference.

SD+ Yield significantly increased through substitution of animal traction for manual cultivation.
SD– Yield significantly decreased when animal traction replaced manual cultivation.
INSEE Institut National de la Statistique et des Etudes Economiques.
BDPA Bureau pour le Développement de la Production Agricole.
CFDT Compagnie Française pour le Développement des Fibres Textiles.
ODM Overseas Development Ministry, United Kingdom.
IDESSA Institut des Savanes.
ICRISAT International Crops Research Institute for the Semi-Arid Tropics.
FSU Farming Systems Unit, Semi-Arid Food Grains Research and Development, funded by USAID.

a. Mettrick (1978).
b. Mettrick (1978).
c. Whitney (1981).
d. Barrett and others (1982).
e. McIntire (1981b).
f. Oluoch (1983).
g. Jaeger (1986).
h. Starkey (1981).
i. Crop area cultivated in which increase was through the use of animal traction.
j. Growth in area per person through substitution of animal traction for manual cultivation.
k. A favorable climate is one in which annual rainfall is 750 millimeters or more; an unfavorable climate is one in which annual rainfall is less than 750 millimeters.

Appendix: A Framework for Evaluating the Benefits of Animal Traction

We have already indicated that a pure cross-sectional comparison of hand-hoe and animal-traction farms tends to exaggerate the advantages of animal-draft power. This is because households that use animal traction typically differ from households that use the hand hoe in systematic ways that would make those that use animal traction more productive even in the absence of animal draft. These differences include size of landholding, farming skill, wealth, access to credit, and information that they possess. These differences affect both the decision to adopt the plow and the decision to use other advanced farming techniques. Therefore the measured differences between farm types confounds the animal-traction effects with the effects of unrelated differences in production techniques.

Before-and-after studies of the same farms can overcome some of these confounding effects. Observations made before the adoption of the technique, however, are generally not available. The next best alternative is to undertake a cross-section study with a retrospective element. For some of the more easily recalled variables, changes made in the course of time that are associated with the introduction of animal traction can be brought to light by retrospective questions. The following discussion will highlight the types of information that can be elicited by recall.

One of the areas in which retrospective questioning can be important is the determination whether the transition to animal-drawn plows leads to an aggregate area expansion. Households that use animal traction can be asked to recall whether they expanded their area of cultivation after the change in power source. Farms reporting an expansion in area must be probed further to determine whether the additional area came from fallow or reclaimed land, displacement of tenants, or rental of land from households that use the hand hoe.

An examination of yield effects requires further detailed questioning, both current and retrospective. Specifically, detailed probing is required as to differences in the quality of tillage on hand-hoe farms and animal-draft farms. Retrospective questions must be designed to determine the quality of tillage before the adoption of the plow—direct seeding, superficial tillage with a hoe, or intensive tillage with a hoe. Farmers who use animal traction should also be asked to recall whether they noticed any increases in yield when they changed tillage quality, such as in the move from direct seeding to intensive cultivation with the hoe.

It is also important to determine whether in the cross-section sample there are any differences in the quality of tillage of farmers who use the hoe and those who use animal draft. Hoe-using farmers should be questioned to see whether they practice direct seeding, superficial tillage, or intensive tillage. To determine the quality of tillage for animal-draft farms we need to know the type of plow used (scratch plow, moldboard plow, or other), the depth of plowing (line tracing, shallow plowing, or deep plowing) and the crops and plots for which plows are used (all crops and all fields, only valley-bottom fields for rice, and so on).

Cross-section studies that include such retrospective questions can overcome the attribution problem to a large extent. First, differences in soil fertility can be accounted for with information on the frequency with which different plots are cultivated, the proportion of fallow land brought under cultivation by households using animal traction, and the intensity of manuring systems. Second, the level of use of other inputs—such as fertilizers, pesticides, and irrigation—on farms with and without animal traction ought to be identified and accounted for in the attribution of yield differences.

Effects on labor use can be obtained by comparing farms with and farms without animal traction. In this comparison it is important to distinguish between labor requirements of crop production *per farm* and *per hectare.* The latter is a measure of direct labor input for crop production, such as land preparation, weeding, harvesting, and so on, while the former includes overhead labor, such as animal care and maintenance and manure production. Measurement of labor input requires close monitoring of the households over the cropping cycle and cannot usually be obtained by recall. In a single visit, however, information on the direction of change in labor use can be obtained through retrospective questions. Households that use animal traction, for example, could be asked several questions to find out whether their labor requirements per hectare for land preparation increased, decreased, or remained the same after they switched from the hoe to the plow.

Finally, information on the effects of animal traction on livestock raising and improvements in animal-husbandry skills can be obtained by recall. This involves questions about whether animal traction complements or conflicts with other livestock-raising activities in requirements for pasture, fodder, or labor time.

Notes

1. There are, of course, specific tillage-quality trials in which, for example, the depth of plowing is varied using the same tool.

2. Note that yields may also be increased by virtue of the weed-control effect of tillage. In tillage experiments, however, the level of weed control is typically held constant.

3. Charreau (1974) reports for semiarid Senegal that a delay of one week in seeding leads to a substantial decline in yield per hectare (see chapter 4).

4. The amount of labor required for land preparation is reduced only if the agricultural intensity or the type of soil is such that tillage is required prior to seeding. A shifting cultivator, for example, does not require tillage of the land, which is soft enough for direct seeding. In contrast, permanent cultivators require several hundred hours of land preparation. Transition to the plow would reduce the labor requirement in the latter instance but not in the former.

5. Farming intensity = (number of crop cycles a year × number of years of cultivation × 100) ÷ (number of years of cultivation + number of years of fallow). See chapter 2 for details.

Chapter 8

Conceptual Issues in the Design of Animal-Traction Projects

The mode of research inquiry of the preceding chapters must now give way to prescriptions for ways of conceptualizing and designing animal traction projects. These prescriptions will be based in part on the analytical framework and the empirical evidence already presented, but they are also derived from the lessons of experience with animal-traction projects to be found in literature more concerned with management, which we have not so far reviewed or which emerged in discussions with animal-traction specialists.

There are three distinct stages in the design of animal-traction programs or projects: a rough determination whether the farming system of a particular area is at a stage at which the use of animal draft is appropriate; a more careful determination of its likely profitability; and identification of the short-run constraints to the transition from the hand hoe that the project should attack. Traditionally, in animal-traction projects great emphasis has been placed on stage three and little or none on the first two stages. We consider this to be the main reason for the poor performance of many animal-draft projects in Sub-Saharan Africa.

Rather than provide a detailed guide for project design and management, we shall identify the information required and the sources from which it can be obtained for each stage in the design process. Table 8-1 summarizes the principal requirements.

The Farming System and the New Technology

In the preceding chapters we have demonstrated the extent to which the introduction and development of animal traction depends upon the

Table 8-1. *A General Checklist for Designing Animal-Traction Programs or Projects in Sub-Saharan Africa*

Item	*Goal or thing to be learned*	*Institution and evaluation as source of data*
	Knowledge of farming system	
Definition of regions of mechanization	Mapping of the country	
Is the use of animal traction possible?	Availability of cattle	Livestock service, often poor
Level of infrastructure	Density of population	Population census, variable
Constraints on and opportunities for mechanization	Physical conditions, conditions of market and extension, general socioeconomic conditions	Universities, generally good, and research centers, generally poor
Basis for on-farm testing and monitoring	Types and characteristics of farms in the region	Universities and research centers; locally improving
Operations to mechanize testing of agricultural equipment	Identification of operations for a specific type of farm in a given region in a given period	Universities and research centers for observations; generally poor use made of available means of research
Mechanization and testing techniques in general	Identification of other techniques, useful or not, for linkage with mechanization	Extension services for on-farm testing; overemphasis on package approach and on package studies
	Assessment of market for meat of draft animals	
	Infrastructure, credit, and market	
	Clear definition of credits, possible subsidies, and market policies before implementation of project	Government; efforts not sufficiently diversified by region and farm

particular farming system. Familiarity with the farming system and its evolution makes it possible to address the first two questions discussed above.

Is Animal Traction Appropriate in a Given Region?

In a particular region the appropriateness of animal traction depends on the farming intensity, agroclimatic constraints, and the availability

Item	Goal or thing to be learned	Institution and evaluation as source of data
	Specific components of project	
Efficiency of veterinary and livestock services	Support for animal traction	Ministry in charge; often poor or conflicting efforts
Training of cattle	Introduction of animal traction	Agricultural extension service; well known
Blacksmiths, craftsmen, training, credit for tools, and the supply of spare parts	Support for mechanization	Ministry in charge, well known
Mechanization and extension	Establishment of links, if they would be useful	Ministry in charge for training and evaluation of extension agents
	National production of equipment	
	Service to agriculture	Ministry in charge; poor perception of real needs and market potential
	Interaction among components of program	
	Effective coordination	Government; too much emphasis on vast integrated projects rather than coordination of services
	Flexibility of programs as consequence of monitoring	
	New times, new region, new farms, new activities	Monitoring services; few; long delays in taking decisions on basis of results of monitoring

of cattle. The relation between farming intensity and the plow was discussed in detail in chapter 2, the principal conclusion of which was that the transition from the digging stick and the hand hoe to the animal-drawn plow would not be made in forest-fallow and bush-fallow systems because of the large labor requirements for destumping and leveling. Destumping requirements are minimal by the time the grass-fallow stage is reached, and it is here that animal power becomes the economically dominant technology. A region-by-region estimate of farming intensity can be obtained from statistics on population density and arable land area. Other things being equal, farming intensities are higher in areas in which less arable land per capita is available. This information is generally easy to obtain—albeit of variable quality—and can be used to determine areas within a country or a region that should be examined further.

The next step is to determine the agroclimatic suitability of intensive production of field crops and therefore the potential for animal traction (see chapter 4). Much of the humid lowlands and some of the highland areas do not have a comparative advantage in intensive production of field crops and therefore need not be examined further. The semiarid zones with high rainfall are most suitable for intensive farming and draft power. Within an agroclimatic zone specific pockets in which animal power could most profitably be promoted need to be determined. These areas are identified by soil type, location on the toposequence, and length of growing period. Animal power for plowing is adopted more readily in areas with heavier soils on lower portions of the toposequence or where growing periods are fairly long. Information on soils, toposequence, and length of growing period can easily be obtained from cartographic, soil, and agronomic studies.

Finally, whether cattle are easily available in the location needs to be determined. If they are not, what are the constraints, such as trypanosomiasis, lack of fodder, and so on, to importing and keeping cattle in the area? Some of the constraints, such as poor veterinary service, can be overcome by the project. Livestock censuses have been conducted in most countries of Africa and do provide a rough estimate of numbers of cattle. Although existing livestock censuses are of poor quality they are usually sufficient for the purpose of determining whether cattle are available in an area.

How Profitable is Animal Traction in a Given Location?

The aforementioned indicators show whether conditions in a particular area are conducive to the introduction of animal traction, but not how profitable it could be. Private and social profitability of animal-draft power is dependent on infrastructure, access to markets, the availability of land, crops grown, and operations mechanized.

Association between Profitability and Crops Grown

As discussed at length in chapter 4, the profitability of switching to the plow is crop-specific. The yield response to tillage is extremely high for rice, so the mechanization of land preparation for rice can be expected to be profitable. Mechanization of land preparation can be profitable even where the yield response is relatively small if either the price of the crop is high or the transition to animal power leads to the cultivation of a larger area. The yield response to tillage is low for the cash crops cotton and groundnut, for example, but animal-draft power is being used for both these crops; the use of animal-drawn plows and other equipment often leads to the cultivation of larger acreages.

Markets and Transport Infrastructure

The existence of market demand for the output and of transport facilities for carrying the output to the market are, of course, necessary conditions for profitable transition to animal-draft power. Most of the available information on markets and transport infrastructure for a particular location is not well documented, but it is easy to obtain reliable information through field visits.

Market opportunities for the meat output of draft animals must also be assessed. Proximity to large urban centers is, of course, an advantage. These market opportunities will codetermine the profitability of animal traction and the age to which to use the draft animals before having them slaughtered.

It is equally important to assess the existence and proper functioning of the market for seed, fertilizer, and other complementary inputs. The farmer should be confident about obtaining a supply adequate to enable him to consider investment in animal traction. Again, information on input markets is most easily obtained through field visits to the area being considered for an animal-traction project. Among the information collected particular attention should be paid to the effects of government intervention on the free flow of inputs.

Availability of Credit

The adoption of animal traction usually represents a substantial investment. Table 8-2 shows that in three projects the investment level with carts ranged from 2.4 to 2.7 times average annual money incomes of the farmers. In areas in which animal traction is feasible, oxen and equipment can often be purchased out of savings in the form of cattle and other livestock. In some instances it has also been possible to purchase oxen and equipment out of nonagricultural earnings from urban jobs, migrant labor work in plantations or mines, cottage industry, and trade. Even though there are many instances of such savings, the number of farmers in any given location with adequate savings is typically low. Credit is frequently required for the introduction of this equipment and animals, especially in the initial stages of the use of animal traction. Access to credit would also prevent the early stagnation of the adoption of animal power, as has happened in several countries, especially in East Africa.

Table 8-2. *Cost of Equipment in Relation to Income in Some Recent Instances of Transition to Animal Traction*

Instance of transition	Average agricultural money income per farm	Various costs: Team	Various costs: Land-preparation equipment	Various costs: Carts	Ratio of investment level to income: With cart	Ratio of investment level to income: Without cart
			1977 CFAF			
Introduction of traction by donkeys in a central area of Burkina Faso, with manual cultivation of some cereal surplus[a]	35,000	15,000	25,000 (hoe)	47,000	2.5	1.1
Introduction of traction by oxen in a groundnut area of Senegal that already had traction by donkeys or horses[b]	70,000	90,000	45,000 (various)	59,000	2.7	1.9
			1980 CFAF			
Introduction of traction by oxen in an area of manual cotton cultivation receiving heavy subsidies in northern Côte d'Ivoire[c]	130,000	140,000	80,000 (various)	90,000	2.4	1.7

a. *Source:* Delgado (1979b).
b. *Source:* Sleeper (1979).
c. *Source:* Bigot (1981).

There is a vast literature on the appropriate design of credit systems for agricultural equipment; see Sargent and others (1981) and Munzinger (1982) for recent reviews of the subject. We are more interested here in obtaining information on credit systems. In general, situations in which credit is now being provided for oxen and equipment are few, and it will therefore be difficult to study the operation and effectiveness of credit systems for promoting draft power. We do, however, find that most countries operate a credit system for tractor cultivation, and an evaluation of this scheme might give some indication of the possibilities for a similar scheme for animal power. Information on tractor credit can be obtained from field visits and from financial institutions involved in lending for agriculture, such as the Agricultural Development Bank. Studies of the effects of credit on the adoption of new technology are generally available at local research centers and extension organizations.

In designing credit systems for animal traction the following general rules that emerge from the literature are useful:

- The introduction of credit facilities does not bring about rapid adoption in situations in which the rate of return on the investment is questionable.
- It is generally preferable for credit to be made available in kind, then recovered in the form of produce.
- The learning period for farmers to take full advantage of animal traction is typically about three years. Repayment schedules must take into account that full benefits will not be realized in less time than this.
- It is typically difficult to channel credit to small farmers. High levels of subsidy on credit further increase the difficulty, because wealthy borrowers will find it more profitable to apply for it whether they need it or not; subsidies on credit should therefore be minimal.

Access to Land

In projects that promote the use of draft animals it is assumed that farmers will have steady, guaranteed access to land and therefore can expand the area under cultivation. It is therefore assumed further that the adoption of animal-draft power will bring about an expansion of cultivated area in the aggregate as well as on the farms that adopt animal traction. As we have already observed, however, the use of animal traction may or may not lead to an aggregate area expansion where access to fallow or uncultivated land is not easy to obtain. Where land is scarce, farmers that use animal traction may expand their area by displacing farmers who still use the hand hoe, who are typically poorer. There are two determinants of whether the introduction of animal-draft power in a given area will lead to expansion in aggregate cultivated area: the availability of uncultivated arable land per capita, and the land-tenure system that is in effect.

Where uncultivated land is easily available most private area expansion will be into that land. Even where uncultivated land is available, however, it may already have been appropriated, for institutional arrangements for access to land become more and more restrictive as population density and improved access to markets make land scarce. The extent to which land already appropriated is available depends on the land-tenure system, in particular whether land rentals, sales, or other transfers are barred. This information cannot be obtained from secondary data but only from socioeconomic or anthropological studies of the issue or through careful questioning during extended field visits. This information is essential to determination of the aggregate welfare of a society where an animal-draft project is proposed.

Essential Components of an Animal-Traction Project

Once it has been determined that the promotion of animal-draft power in a given area is feasible and profitable, the next problem is to identify the short-run bottlenecks in the switch from the hand hoe. Animal-traction projects are generally designed with an eye to releasing these bottlenecks, such as in veterinary care, animal training, and the availability of credit. Readers unfamiliar with the design of specific components to be included in animal-traction projects are referred to Munzinger (1982).

Veterinary and Livestock Services

Veterinary and livestock services must provide two distinct services: prophylactic services, usually for the entire regional herd, and specific curative treatments for draft animals. A third component, extension efforts designed to promote better feeding of animals and the complementarity between meat, dairy, and draft cattle, is often specific to animal-traction projects.

GENERAL PROPHYLAXIS. A prerequisite for animal-draft projects is that the principal risks to animal health should have been eliminated or at least brought under control. Cattle diseases, such as rinderpest, foot-and-mouth disease, and cattle plague, can be controlled by vaccination. Tick-borne diseases can be controlled by cattle dips (see Munzinger 1982, 11–123, for an overview of health risks and diseases of draft animals). Monitoring of cattle movements and thereby the control of epidemics is also an essential part of this service. Before designing an animal-traction project, therefore, one needs to be sure of the availability of immunization services provided by the veterinary service, an assessment that must be made locally. In our field visits we found that general prophylaxis for cattle and other livestock was typically administered by livestock service and was free in the majority of cases. More intensive questioning is required, however, for the assessment of the quality and mobility of these services in the planning of animal-traction projects (see table 8-3).

SPECIFIC CURATIVE TREATMENT FOR DRAFT ANIMALS. Draft animals are often the most valuable animals of a region, but the incidence of injury to them is higher than to other cattle, and they may be more susceptible to certain kinds of disease. The benefits of a good veterinary service are therefore high, especially where the animals have been obtained on credit or where they are insured against death. The introduction of animal traction therefore calls for modifications in existing veterinary services, and these may not be easy to achieve. Promoting draft power is often not part of the mandate of the ministry of livestock or the livestock department. Special treatment for draft animals requested by officials outside the department may therefore not be honored. In the absence of an integrated policy for the promotion of animal traction, there is no mechanism that can mediate between

Table 8-3. *The Availability of Veterinary Services in the Villages Visited by the Authors in 1983–84*

Item	*Curative*	*Preventive*
Total number of instances	48	48
Number of instances in which veterinary service was available	42	43
Number of instances in which the service was free	13	28
Distance to the service (kilometers)		
0–4	16	22
5–9	3	2
10–14	8	7
15 or more	15	12

agricultural services and veterinary services to mobilize existing resources for providing specific veterinary care to draft animals.

EXTENSION EFFORTS TO PROMOTE COMPLEMENTARITY. Livestock services and their projects are typically designed and implemented separately from other agricultural pursuits, and the promotion of animal draft power is not often considered to be a component of such projects. Extension efforts, however, should promote the complementarity between meat, dairy, and draft functions of cattle, especially where markets for these products are well developed. Experience in Sub-Saharan Africa has shown that after several years of use draft oxen can often—but not always—be sold for slaughter at prices higher than meat cattle of similar age.

The use of cows for draft has been common in Senegal, Botswana, and Mozambique. Experimental trials for introducing cows as draft animals are under way in south central Senegal, in northern Cameroon, and at the International Livestock Center for Africa (ILCA) in Ethiopia. The successful integration of crop and livestock activities in mixed-farming households will benefit from complementary rather than contradictory extension advice from the concerned departments or ministries.

APPROPRIATE INTEGRATION OF THE CROP AND LIVESTOCK SUBSECTORS. First, care should be taken not to link livestock raising and animal traction in situations where neither is likely to develop satisfactorily—that is, in areas that are too humid or where the technical and agricultural policy prerequisites are lacking. The success of this integration depends on the appropriateness of draft power for cultivation. The use of animals for draft can provide an impetus for production of high-quality meat. The work performed by the animals, the special feeds they receive, and the veterinary care that is provided for them can all contribute to better meat quality. The production of meat can, in some instances, be the primary objective, with draft power a second objective (Lhoste 1983). The introduction of animal traction brings about changes in the structure of the livestock subsector (Sleeper 1979). New possibilities for contact between herders and farmers emerge because of the demand for young oxen for training, the possibility of finishing off cattle, and so on. This development should be encouraged, because it ensures the supply of oxen and of premium beef. Such integration can sometimes allow herdsmen to pass from a precarious situation of exclusive livestock raising to one that includes cropping. In many parts of Sub-Saharan Africa such mixed farming is replacing exclusive herding activities.

Training Animals and Farmers

The methods of training animals for draft in Sub-Saharan Africa are well documented according to animal type and species, as well as according to the techniques for different types of yoke, harness, and equipment (FAO-CEEMAT 1972, Munzinger 1982). Suffice it to say here that there is no one method which is applicable to all situations and that a technique suited to one place cannot automatically be transferred to another, given the diversity of animals and agroclimatic conditions in which they have to work. Here we shall discuss certain aspects of the organization of training that have implications for its success.

One important point of concern is whether the training centers should be located outside the village and the specific rural setting in which the animals are to be used. Despite the scale and locational advantage of a central training facility, it is frequently found that such centers have not brought about any widespread diffusion of animal traction. The use of pilot farms as training centers has not worked either. Such centers have used technologies, such as sophisticated animal shelter and feeding arrangements, that were far beyond the reach of the average farmer.

Small, regional centers organized along simpler lines can be more useful in the initial stages of the use of animal traction, especially if the farmers and oxen are trained under conditions similar to those on their own farms. Investments in training centers will be all the more cost-effective if, after having served in the introduction of animal traction, they are then used as centers for training in the use of new equipment or new practices in livestock raising or crop growing. A number of national extension services and technical assistance institutions have wide experience in this field, and the design of an appropriate training system is not a serious problem.

When the training centers are used for extension of a wide variety of crop and livestock technologies, care should be taken to see that there is no unnecessary link between the training of farmers to use draft power and the diffusion of other technologies. The introduction of an overabundance of new techniques during training activities leads to unmanageable sessions and sometimes actually impedes the transfer of basic animal-traction skills.

We do not, however, advocate that animal-draft training be offered in isolation from all other activity. We do recommend that crop and livestock extension agents be present at training sessions to give advice on the location-specific conditions to which the training is relevant. These agents could also counsel the farmer in ways of acquiring and using draft animals and equipment. It is therefore essential that extension personnel involved in this training should be given the analytical tools that will enable them to give advice on the pertinent choices among the generally wide range of possible technical actions.

Training of Blacksmiths and Other Craftsmen and Spare Parts Supply

The distribution of equipment will create a demand for maintenance services and spare parts. This problem has been attacked in a number of ways; by far the most effective has been through the training of blacksmiths. We are not aware of training programs for carpenters and leatherworkers, although in Asia these craftsmen produce yokes and harnesses. Apart from yokes and sleds, we have found virtually no wood-based tools or carts in our field visits in Sub-Saharan Africa.

The establishment of central workshops to which the farmers must come to have equipment repaired or to procure spare parts, however, has proven to be insufficient. The real answer is to train blacksmiths who will be widely dispersed in the villages and integrated into them. No public or artisanal organization in a semiurban or urban setting can supply this service in the same way.

Traditional artisans do exist in most rural areas but the demands on their services change as the farmers acquire new technologies and new equipment. Animal-traction projects should, therefore, include retraining programs for these artisans so that they can repair animal draft equipment.

Experience with blacksmith training in Sub-Saharan Africa has shown that there is a good potential for village repair and production using scrap metal and a minimum of imported or new parts. The upkeep of animal-draft equipment has not been a serious problem in most countries.

Well-trained blacksmiths can be useful, not only in repairing equipment, but also as auxiliaries in the establishment of plants for the assembly or manufacture of equipment. This kind of evolution took place in the developed countries with the introduction of mechanized agriculture. As Ruttan (1980, 79) observed, "In the area of mechanical technology the private sector has been remarkably successful in generating a continuous stream of new technology. The classic pattern has been the invention of new machines or equipment by farmers, blacksmiths or mechanics followed by improvements in design and performance by the R&D units of the farm equipment companies." Recent examples from Sub-Saharan Africa—the manufacture of carts in Madagascar and plows for local use in Guinea—indicate that such evolution is possible here also. This presupposes, however, that any industrial manufacturing that may be present or may emerge will complement rather than compete with the activities of the blacksmiths. Country-specific conditions should be examined to determine that this is indeed so.

Domestic and Regional Production of Equipment

Mechanical technology is sensitive to agroclimatic factors such as soils, terrain, and rainfall regimes and to economic factors such as the availability of capital and materials and the size of farms. When there is a divergence in either environmental or economic conditions direct transfer of mechanical technology is limited—hence the case for local manufacture of equipment. Several plants for the manufacture and assembly of equipment have been started during the past two decades in most countries of Sub-Saharan Africa. Their production activities cover a wide range, from the manufacture of tools for manual cultivation and equipment for animal traction to the manufacture of motorized equipment.

In most instances, however, the agricultural machinery industry has not displayed the ability to survive and to sustain operations without high levels of government support and subsidy. This seems surprising, since broadly based agricultural development can and should give rise to a growing demand for a widening range of simple, inexpensive farm equipment, and this demand fosters the growth of a local farm-machinery industry (Johnston, 1981, 30). Why has local manufacture of farm equipment not succeeded in Sub-Saharan Africa? This may be in part a consequence of a top-heavy approach: the manufacture of machinery was established, with government support, in large factories rather than by small entrepreneurs.

A review of the development of the agricultural machinery industry in the United States (Evenson 1982) and in selected developing countries (Binswanger 1984) shows that in the early phases of mechanization, invention of equipment and adaptation of it to local conditions are done by small manufacturers or workshops in close association with farmers. The emergence of a diversified machinery industry out of small shops in Punjab, India, and the power-tiller industry in Thailand and the Philippines all followed a similar pattern. In the United States, the contribution of large corporations gradually increased but continues to be most important in the area of engineering optimization. It was only around the beginning of the twentieth century, for example, that the plow industry in the United States became consolidated, the large firms, such as John Deere, purchasing the patents and assets of small firms as they expanded (Binswanger 1984).

In Africa, on the other hand, individual initiative was suppressed under the pressure of national considerations shored up by diverse technical assistance. The result was fairly large firms producing equipment not adequately matched to local conditions. The consequence of the establishment of a national farm-machinery industry may well have been to drive local craftsmen out of the machinery-development market. These craftsmen can produce equipment at prices 30–50 percent lower than the large industries but do not have as easy access to raw materials and equipment because of import restrictions. The result is a disruption in the subsector that has traditionally fulfilled an important function in the development of machinery.

The second major problem was that the need to adapt to local conditions was confused with the need to establish fairly large national factories. The creation of national plants for assembly and production has not brought any amelioration of the problems that used to cause friction between local farmers and foreign suppliers of equipment. The national enterprises are now facing the same problems that foreign suppliers used to have. When foreign manufacturers were supplying equipment, field trials were conducted and design changes were made in the tropics but production took place overseas. This caused the emergence in Europe of a whole industry that specialized in the production of equipment for Sub-Saharan Africa. This industry shut down its activities when a host of national factories were created during the 1960s. The failure of the national machinery industry is the result of inadequate capital and the difficulties of obtaining supplies from abroad, continued production of obsolete equipment, and failure to adapt to local farm conditions.

In designing animal-traction projects one must, therefore, consider the following: In the initial stages of the introduction of new equipment efforts should be made to ensure the availability of reliable supplies of appropriate equipment from abroad rather than to manufacture equipment within the country. Government policy should encourage the participation of small entrepreneurs in the manufacture of equipment. As the demand for this equipment increases small entrepreneurs will begin local manufacture, using foreign designs that have been adapted to local conditions. The availability of foreign exchange for tools, steel, and parts is particularly important. Attempts to supersede individual initiative by establishing public enterprises or by subsidizing and protecting large private enterprises, including foreign-owned factories, should be discouraged. These enterprises do not have a comparative advantage in the manufacture of fairly simple equipment, they require high levels of subsidy, and they do not capture the innovations and adaptations of farmers. The supply of repair services and spare parts should not be centralized; they should be provided by local blacksmiths, artisans, and small entrepreneurs.

Part Three

Motorized Agriculture

Chapter 9

The Selective Pattern of Agricultural Mechanization

Our discussion to this point has been concentrated on the transition from hand to animal power and the tractor for the tillage operation. There are, however, other operations that are transferred to animal or mechanical power either concurrently with or before the tillage operation. Sole concern with the mechanization of tillage obscures the successful switch in sources of power for operations such as transport, milling, and pumping. In much of Sub-Saharan Africa transport and milling were mechanized long before the mechanization of tillage. In East and South Asia the first operation to be motorized was water pumping, not land tillage.

Agricultural operations can be classified by their use of power in relation to human judgment; the first operations to be mechanized are the ones that use a large amount of power and little or no human judgment. Operations in which human judgment is important are mechanized only when wages begin to rise. The selective pattern of mechanization discussed here was the same in most countries during the early stages of mechanization, but the sequence in which power-intensive operations are mechanized depends on the economywide factor endowments. Where land-to-labor ratios are low, for example, mechanical pumping comes before mechanical tillage.

Power Intensity versus Control Intensity

Agricultural operations can be grouped according to the relative intensity with which they require power, or energy, in relation to the control functions of the human mind, or judgment. Operations such as land preparation, transport, milling, grinding, and threshing are power-intensive, while weeding, sifting, winnowing, and fruit harvesting, for example, are control-intensive operations. The shift of the source of power from human to animal to mechanical power is dependent on the level of power intensity and control intensity of the operation. New sources of power are always used first for power-intensive operations, regardless of the stage of mechanization. It is for these select operations that the use of the new power source has the greatest comparative advantage. Furthermore, it appears that the mechanization of power-intensive operations is less dependent on the price of labor than is the mechanization of control-intensive ones—that is, it often pays to move to a higher stage of mechanization in power-intensive operations, even at low wages, when mechanization of control-intensive operations is not profitable.

Table 9-1 offers a comparison of operations according to their power intensity and control intensity and the sequence of their transfer to the new power source. The stationary operations, such as milling, grinding, threshing, and water lifting, are among the first to be transferred to a new source of power. The transport operation is the other mobile operation besides tillage that is quickly shifted to animal and mechanical power. It should be noted that the constraints imposed by low farming intensity on the adoption of the plow are not applicable in the case of milling, grinding, and transport. These operations are mechanized even under low farming intensities because of the enormous savings in

Table 9-1. *Comparison of Agricultural Operations According to Their Power and Control Intensity*

Nature of operation and source of power	*Low control intensity, high power intensity*	*Intermediate intensity of power and control*	*High control intensity, low power intensity*
Stationary operations	Grinding, milling, crushing Water lifting Threshing, wood cutting		Sifting, winnowing
Mobile operations	Transport	Harvesting root crops Harvesting grain crops	Weeding and harvesting tea, coffee, and apples
	Primary tillage	Secondary tillage and interculture	Seeding

labor as a result of a switch away from human power and the ease of establishing rental markets for the use of this equipment, which does not suffer from the timeliness problem faced in the tillage operation. Operations, such as harvesting, with intermediate power and control intensities, are mechanized only after the mechanization of labor-intensive operations and are more sensitive to the level of wages. The transition away from hand power would take place in the face of rising wages. This selective nature of the adoption of a new sourcc of power—that is, the coexistence of hand, animal, and motor power—was also observed in Europe and the United States.

The Mechanization of Stationary Operations

Power-intensive operations such as milling, grinding, and pumping water are often the first operations to be mechanized with the use of stationary motors.

Milling, Pounding, Grinding, and Crushing

Postharvest processing operations are extremely labor-intensive and tedious to perform. Miracle (1967) reported, for instance, that to grind a week's supply of maize meal—thirty pounds—by hand would take from eight to fifteen hours. The same operation would take half an hour with a hand mill and perhaps not more than ten minutes with a motorized mill. The same is true of dehusking rice, crushing sugarcane, grinding groundnuts, and ginning cotton. In most parts of the world these operations have been transferred to stationary machines powered by water, wind, steam, and—more recently—internal combustion engines or electricity.

Water was first used for milling, pounding, and grinding in the first century B.C. in China, and its use for these purposes was fairly widespread between the second and fourth centuries A.D. Water-powered milling had also been adopted in all corners of Europe by the twelfth century A.D. Wind power was used concurrently for milling and lifting water in Europe. With the advent of steam power, mills were transferred to this source of power in the nineteenth and early twentieth centuries in both Europe and the United States. By the outbreak of the U.S. Civil War, steam power had almost completely replaced horses and oxen for powering sugar and rice mills and to turn cotton gins. With steam power, three men and a cotton gin could remove the seeds from 1,000 to 4,500 pounds of cotton a day, which was about a hundred times the amount they could gin without steam power (Hurt 1982, 101).

In South Asia animals have long driven Persian wheels, sugarcane crushers, and oil crushers, but the animals used in these operations are gradually being replaced by diesel and electric engines. In India in 1973 the number of stationary engines for power-intensive operations was about twenty times the number of tractors (India 1975 and 1976). In all of Asia mechanical milling of large traded quantities of rice had already been introduced in the late nineteenth century, usually by steam engines, later by internal combustion engines. Smaller rice mills have swept across Asia since the 1950s, and it is hard to find villages where rice is still pounded by hand.

Mechanical mills were introduced in Africa after World War I and spread rapidly through the continent. There is documentation of the earlier existence of water mills in Angola and Kenya (Manners 1962, Jones 1959). Jones (1959) and Miracle (1967) have reported widespread use of mechanical mills, both hand and motorized, across Sub-Saharan Africa. Miracle states that he has seen or has records of mills of one sort or another throughout the southern parts of Côte d'Ivoire, Ghana, Togo, Benin, Nigeria, Cameroon, Zaïre, Kenya, Tanzania, Zambia, Malawi, Zimbabwe, and Mozambique.

During our field visits we found mechanical mills used in thirty-four of the fifty-five locations visited. Figure 9-1 shows the frequency of occurrence of mechanical mills, by farming system. The adoption of the mechanical mill often precedes the use of the plow and parallels use of the plow with increases in farming intensity; the frequency of use of the plow is shown in figure 2-3. The important point to note here is that low intensity of farming is not a constraint to the adoption of mechanized mills. This is mainly because the labor input required for milling is independent of the intensity of farming, and mills are rarely owned by the individual households who use them. The service is provided on a charge-per-unit basis by private entrepreneurs or village cooperatives. Since mills do not face any of the timeliness problems associated with plows, efficient rental markets are easily established and the cost of the equipment is spread over many users.

The Threshing Operation

Mechanical threshers were quickly adopted wherever harvested volumes were large and the wage rates high. In this connection it is useful to examine technological change in threshing operation in early U.S. agriculture as documented by Hurt (1982). Until 1850 colonial farm-

Figure 9-1. *Farming Intensity and Mill Use*

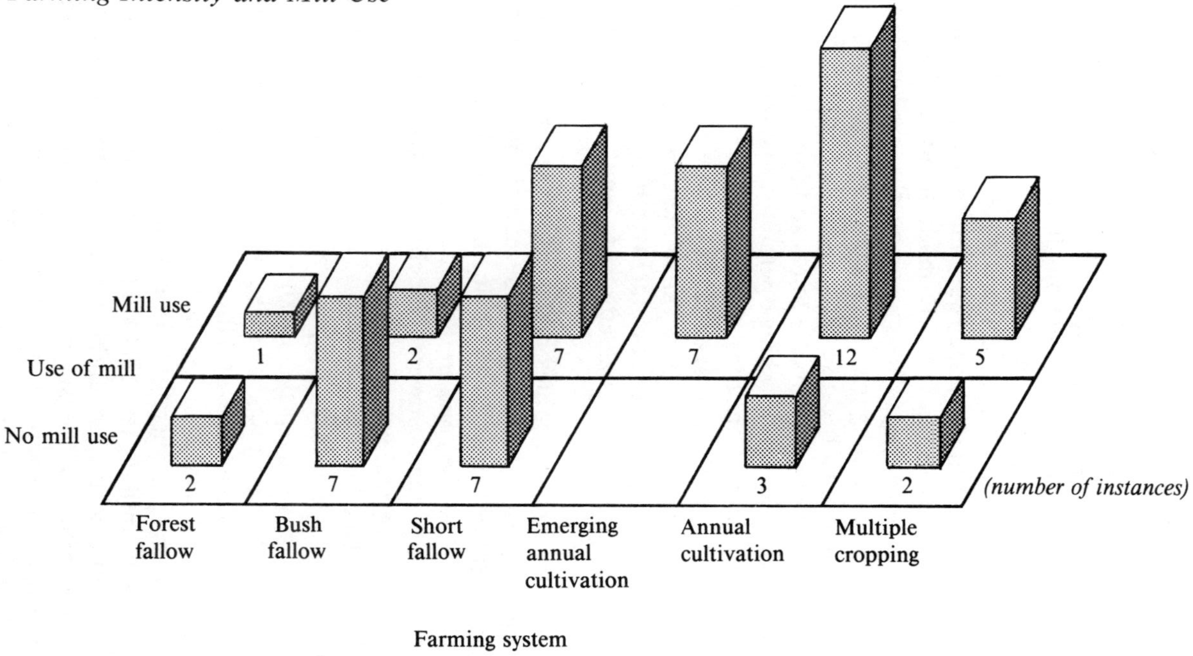

ers, particularly those in New England, used the hand-held flail to thresh grain from the heads. The flail consisted of a short wooden club attached to a long handle by means of a piece of leather. In the midwestern states, where harvested quantities were much larger, farmers used oxen or horses to tread the grain from the heads.

The first horse-powered threshing machine was patented in the United States in 1791, but it was only between 1820 and 1830 that a number of small, simple, inexpensive, and locally made hand- and horse-powered threshing machines began to appear on the market. Most farmers, however, found threshing with the flail to be cheaper than investing in a threshing machine, because the work could be done during the winter, when there was an abundance of cheap farm labor. It was only with the advent of contract threshing operations that mechanical threshing became profitable. Threshing machines were owned by an entrepreneur, who sent a thresher with an itinerant crew from farm to farm. Although contract threshing imposed an immediate cash expense on the farmer, it did free him from the capital investment necessary to purchase a threshing machine and enabled him to get the crops to the market before prices fell.

Steam-powered threshing machines preceded steam-powered tractors by more than ten years. In less than a decade steam had almost entirely replaced horses for power. By 1880 the Bureau of the Census estimated that 80 percent of the grain in the principal wheat-producing states was threshed by steam-powered machines. Steam-powered threshing machines were followed in the 1930s by tractor-powered harvester-combines. The adoption of threshers in Europe followed the same pattern. In 1950 agriculture in Japan was in the early stages of mechanization, with many small pedal threshers and power tillers. This happened in response to rising wages caused by a rapidly growing industrial sector. By 1960 there were one thresher per 2.5 hectares and one power tiller per 12 hectares in Japan (Herdt 1983).

The use of mechanical threshers did not emerge in South Asia and parts of Southeast Asia until the late 1960s. This is not surprising, since wages were low, capital costs high, and harvested volumes small. But where the green revolution raised wages and increased harvested volumes, small threshers were rapidly adopted in Indian Punjab, the Philippines, and Central Thailand as soon as efficient designs were available. By the early 1980s the new threshers were penetrating into other South Asian regions (Walker and Kshirsagar 1981). As in the United States in the nineteenth century, these threshers are owned by private entrepreneurs who thresh on a contract-hire basis.

Mechanical threshers are still rare in African agriculture. Threshers are not yet profitable in Africa because the harvested quantities per person are small. Therefore there are two conditions under which threshers would be profitable in Africa: improvements in access to markets or seed varieties leading to increases in harvested output and rapidly rising wages, which would increase incentives for adopting labor-saving technology.

Irrigation

Drawing water for small-scale irrigation is extremely labor-intensive, and mechanical devices that can relieve the burden have been used for thousands of years. The counterbalance lift, the oldest and simplest lifting device, has been used for millenia in the Arab Republic of Egypt and India. The Persian wheel, which makes use of animals to supply rotary power, is still used in South Asia. The treadmill, also an ancient device, continues to be used in China. Today electric and diesel engines are commonly used for lifting water in most parts of Asia and parts of North Africa.

The growth of farmer-initiated small scale irrigation systems is closely associated with farming intensity. It is only under conditions of land scarcity that irrigation systems are used for extending cropping to two or three seasons a year or for cultivating marginal lands. In East Asia investments in water control and irrigation preceded all other forms of mechanization. Most of Japan's rice fields were supplied with irrigation facilities by the late nineteenth century. The same thing occurred in Korea and Taiwan before 1940. Initially, irrigation systems were based on gravity flow and foot-operated pumps, but with the availability of electric and diesel engines, power pumps became the first machinery investment for many rice producers. Electric irrigation pumps replaced foot-operated pumps in Japan during the 1920s, long before power tillers were used. By 1966 in Taiwan there was one water pump for every three farms, while there was only one power tiller for every eighteen farms (Herdt 1983). Since the 1950s electric and diesel pumps have spread throughout semiarid India (Engelhardt 1984).

During the twenty-year period 1950–70 the number of diesel pumps in India grew from 26,000 to more than a million. In the same period the number of tractors rose from 8,500 to more than 90,000 (Raj 1973). These figures clearly indicate the importance and rapid adoption of mechanized irrigation systems in land-scarce economies. Small-scale irrigation systems in Sub-Saharan Africa have relied mainly on gravity flow and are closely associated with rice production in the valley bottoms. We are not aware of any rapid increases in mechanized pump irrigation systems. This is to be expected, since in large parts of Africa cultivation has not yet been forced onto marginal lands, and irrigated second crops are uncommon.

The Mechanization of Mobile Operations

Among the mobile operations that could be mechanized are primary tillage, transport, seeding, crop husbandry, and harvesting. Of these primary tillage and transport are extremely power-intensive and are transferred to a new source of power even when wages are low.

Seeding and planting can be done with great precision using animal-drawn and tractor-drawn equipment. The use of such equipment may lead to modest gains in yields and facilitates weeding and interculture operations. When wages are low the use of seeders may be profitable where the additional yield is very valuable because land is scarce or where a market crop is grown and the growing season is extremely short, so that the use of seeders enables the cultivation of larger areas than hand seeding. The use of seeders for groundnuts in Senegal illustrates the second situation.

Weeding and interculture operations are control-intensive, requiring the ability to discriminate between the plant and the weed. Interculture with draft animals requires line sowing and may or may not emerge rapidly after the introduction of the plow. Even where it does emerge, hand weeding is still required within the rows. Only where wages rise to very high levels is intrarow weeding transferred to herbicides.

Harvesting operations are of intermediate to high control intensity and are rarely mechanized in countries in which wages are low. Where harvested volumes are large and wages are high this operation tends to be mechanized rapidly. In Sub-Saharan Africa mechanical harvesting is confined to large-scale commercial farms. Mechanization of this operation is not likely to be profitable in the smallholder sector.

This power-intensive operation is quickly shifted to new mobile sources of power when they become available. Carrying loads was the earliest use of domesticated work animals, even preceding tillage. The donkey was the chief pack animal in Sub-Saharan Africa, especially in the savanna. The camel substitutes for the donkey in the Sahel. Human portage persisted only in the forest zone because of the problem of trypanosomiasis and lack of pasture (Hopkins 1973, 72).

The pioneers of commercial passenger and freight services in Sub-Saharan Africa were Africans, not European settlers. By 1940 French West Africa had about 10,000 motor vehicles, while it had a mere sixteen in 1913. The expansion of motor transport in British West Africa was even faster; during the 1920s twice as many vehicles were entering the Gold Coast as were entering French colonies (Hopkins 1973, 196). The story is similar in East Africa: Sukumaland, Tanzania, was well connected by motor transport as early as 1926, and headloading had completely disappeared (Iliffe 1979, 286). Similar statistics for Uganda, Kenya, and Zambia can be found in Mukwaya (1962), Manners (1962), and Colson (1962).

Our field visit information, presented in figure 9-2, validates the proposition that the transport operation is quickly mechanized. In forty-three of the fifty-five locations it was stated that there was access to motorized transport to the market. In most instances transport to the market was by means of rural bus or taxi. Farmers with large quantities of output could generally rent trucks or tractors for transporting produce on an individual basis. In forty-three of the fifty-five instances, moreover, it was reported that rental markets for transport of produce were prevalent. In most instances it was also reported that bicycles or pack animals were used for field-to-home transport. The use of carts or sledges was reported in parts of Kenya, Tanzania, Botswana, and Burkina Faso, although in none of these locations is use of the cart nearly so widespread as it is in Asia. In general, the use of ox carts or sledges is rare in Sub-Saharan Africa.

There are several plausible explanations for the lack of carts in Africa. First, in precolonial times, when population was sparse, large areas were under bush and forest cover, so the use of carts was infeasible. Second, where it was technically feasible to keep the cart, such as in the savannas, the farming system did not generate a great demand for on-farm transport. Where population density is low cattle are not fed at the stall, so transport of fodder from field to homestead is unnecessary. The fertility of the soil is regenerated by letting the land lie fallow, so there is no demand for transport of manure, compost, or silt. Finally, under subsistence production the harvested quantities are small and can be transported to the homestead easily by head load or donkey and camel portage.

Third, because population was sparse road networks were not well developed and the distances between home and market were greater. Pack animals were better suited to traveling on the narrow tracts and were faster than the carts.

The use of animal-drawn carts in Europe was inhibited by many of the same problems that existed in Africa. In eighteenth-century Spain, for example, pack animals, especially donkeys, were by far the most important means of transport, though ox carts were available and used to a certain extent. Carts became popular all over Europe in the mid eighteenth century when farming systems intensified and the roads had been improved (Hopkins 1973, 74).

Road networks in Africa were improved substantially in the mid twentieth century. These improved roads had an immediate effect on

Figure 9-2. *Farming Intensity and the Mechanization of Transport to Market*

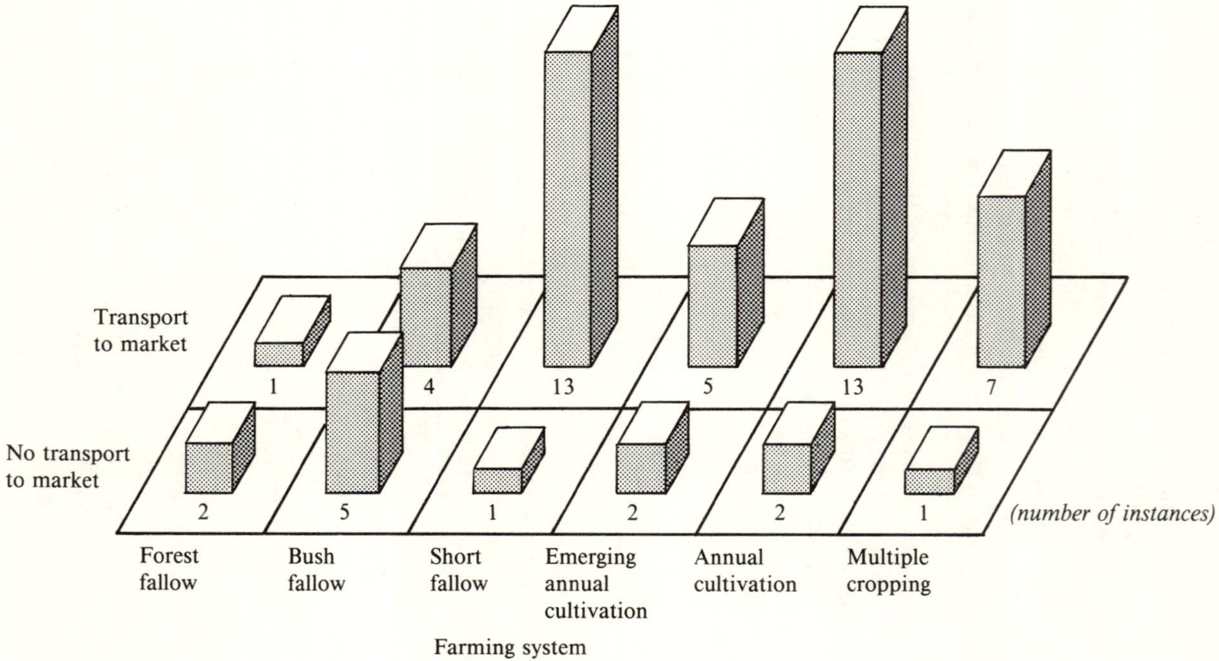

the transport of agricultural products, mainly through rapid expansion of trucks, buses, and bush taxis. The use of animal-drawn carts does not appear to have increased rapidly, however, in spite of improvements in the roads. This may be because it is often cheaper for individual farmers to rent space on a truck or a bus to carry goods to the market as the occasion demands than to buy and maintain a cart. Unlike in densely populated South Asia, the long distances to the market in most of Sub-Saharan Africa imply that slow-moving carts have a comparative disadvantage in relation to motorized vehicles. Note also that transport from field to home and to nearby markets was often shifted from head load or donkey to the bicycle.

Factor Endowments and Agricultural Mechanization

The history of agricultural growth in the developed world illustrates that the rate and direction of technical change are influenced by the land and labor endowments of an economy, by the nonagricultural demand for labor, and by the conditions of demand for final agricultural products. In agriculture the constraints imposed on development by an inelastic supply of labor may be offset by advances in mechanical technologies, while constraints imposed by an inelastic supply of land may be offset by biological technology. This responsiveness of technical change to economywide factors has come to be known as the process of induced innovation (Hayami and Ruttan 1985, Binswanger and Ruttan 1978). The induced-innovation hypothesis becomes clear when agricultural growth in Japan and the United States are compared. Land-scarce Japan emphasized biological, yield-increasing technology supported by heavy investments in irrigation, along with intensive manuring techniques, while the United States emphasized mechanical technology.

The induced-innovation hypothesis should not be taken to imply that there is no future for mechanical technologies in densely populated, land-scarce economies. Indeed, during the late 1960s and 1970s partial mechanization of agriculture could be observed at a time when wages were low— in the Indian Punjab, Thailand, and the Philippines, for example. In land-scarce, low-wage economies mechanization would be seen, but with a selective emphasis on power-intensive operations, and the sequence in which operations were mechanized would differ from those in land-abundant economies. Table 9-2 presents several examples of the use of tractors, by operation, in South Asia.

In both land-scarce and land-abundant economies the power-intensive operations are the ones to be mechanized first. The sequence

Table 9-2. *Use of Tractors in India*

City, state, region	*Range of farm sizes or average size of farm (hectares)*	*Total hours used*	*Use by owners as percentage of total hours used*							
			Tillage	*Irri-gation*	*Thresh-ing*	*Sowing*	*Trans-port*	*Total agri-cultural uses*	*Non-agri-cultural uses*	*Hired out*
Punjab, India[a]	7–10	682	35.9	19.9	11.6	1.3	3.4	72.7	19.8	7.5
	10–20	792	43.6	18.7	9.2	3.3	5.8	80.6	17.7	1.7
	> 20	1,008	49.7	8.3	11.8	3.6	5.1	79.5	18.8	1.8
Punjab[b]										
Region 1	10.6	655	n.a.	n.a.	n.a.	n.a.	n.a.	70.4	26.9	0.0
Region 2	9.5	707	n.a.	n.a.	n.a.	n.a.	n.a.	90.9	9.0	0.1
Region 3	10.9	279	n.a.	n.a.	n.a.	n.a.	n.a.	87.9	12.1	0.0
Region 4	8.3	360	n.a.	n.a.	n.a.	n.a.	n.a.	89.5	9.4	1.1
Region 5	15.5	550	n.a.	n.a.	n.a.	n.a.	n.a.	63.4	33.5	3.1
Haryana[c]	–6–10	(278)	68.6	0.0	12.0	1.0	6.2	87.8	0.0	12.2
	10–14	(407)	70.1	1.0	11.3	0.7	10.1	93.1	0.0	6.9
	14–20	(575)	68.5	6.9	7.9	0.5	9.8	93.4	0.0	6.6
	> 720	(870)	73.7	3.9	11.5	1.5	8.0	98.6	0.0	1.4
Punjab, Pakistan[d]	+0–24	1,019	n.a.	n.a.	n.a.	n.a.	n.a.	n.a.	n.a.	23.6
	24–49	1,273	n.a.	n.a.	n.a.	n.a.	n.a.	n.a.	n.a.	24.7
	+49–73	1,325	n.a.	n.a.	n.a.	n.a.	n.a.	n.a.	n.a.	8.9
	> 73	1,523	n.a.	n.a.	n.a.	n.a.	n.a.	n.a.	n.a.	0.4

Delhi[e]	0–6	375	n.a.	n.a.	n.a.	n.a.	n.a.	n.a.	n.a.	9.1
	6–10	672	n.a.	n.a.	n.a.	n.a.	n.a.	n.a.	n.a.	5.2
	> 10	1,243	n.a.	n.a.	n.a.	n.a.	n.a.	n.a.	n.a.	0.7
Gujarat[f]										
Dascroi (TO)	9.6	655	28.6	0.0	2.7	0.0	18.6	49.8	5.8	44.3
Anand (TO)	7.1	882	15.1	0.0	0.5	0.1	15.7	31.4	9.5	59.1
Dholka (TO)	35.3	861	25.7	0.0	5.9	3.8	20.7	56.1	6.9	37.0
Dascroi (TH)	4.6	(55)	76.0	0.0	12.7	0.0	11.3	100.0	0.0	—
Anand (TH)	3.4	(57)	59.7	0.0	40.3	0.0	0.0	100.0	0.0	—
Maharashtra[g]	41.5	544	(51.6)	(23.2)	n.a.	n.a.	(17.1)	n.a.	n.a.	34.0
Chittoor, Andhra Pradesh[h]	11.0	475	21.9	10.5	2.9	0.0	12.6	47.9	29.3	22.7

n.a. Not available.

— Not applicable.

TO Tractor owners.

TH Users of hired tractors.

Note: Figures in parentheses are not strictly comparable; see the notes in the source studies.

a. Government of Punjab (India) (1974).

b. Kahlon (1975).

c. Sharma (1974).

d. McInerney and Donaldson (1975).

e. Motilal (1971).

f. Desai and Gopinath (1975).

g. Sapre (1969).

h. Narayana (1977).

Source: Binswanger (1978, 48–49).

in which these operations are mechanized, however, differs according to land endowments. In land-scarce economies the pumping operation is generally the first to be mechanized, using diesel and electric pumps. Pumps provide a supply of water for double or triple cropping and allow the expansion of cultivation onto marginal lands. The rapid spread of the use of pumps in Japan, Taiwan, and India has already been discussed. The story is similar for the Philippines. The exhaustion of the land frontier in the 1960s brought about a large demand for small-scale irrigation and the number of pumps grew accordingly, from almost none in 1960 to around 10,000 in 1971 (Barker and others 1973). Mechanical mills, tillage, and transport equipment follow the adoption of pumps. In land-abundant economies, the use of pumps is delayed until conditions of land scarcity emerge; milling and transport operations are the first to be mechanized, followed by tillage where its mechanization is feasible. The threshing operation, although power-intensive, is generally not mechanized where wages are low and harvested volumes are small. Even when land is abundant, therefore, if agricultural production is mainly for subsistence, threshing is the last of the power-intensive operations to be mechanized.

Mechanization of control-intensive operations is more closely associated with the wage rate. In land-scarce economies in which nonagricultural demand for labor is low, operations such as weeding, interculture, and harvesting continue to be done by human and animal power. Cultivators became prevalent in Japan during the late 1950s, when agricultural wages rose in response to rapid postwar industrialization. It was only in the 1970s that rice transplanters and harvest combines began to be used in Japan (Okawa and others 1966). In India and the Philippines, where tractors do tillage and transport, interculture continues to be done by hand and animal-drawn equipment, while harvesting is done only by hand. In land-abundant economies where market opportunities are good and the wage rates are high, many of the control-intensive operations are transferred to mechanical power. U.S. farmers switched from horse-drawn to tractor-drawn cultivators around the late 1920s, when the International Harvester Company introduced the first affordable, efficient row-crop tractor. Tractor-powered harvest combines were adopted widely in the midwestern states during the 1930s (Hurt 1982).

In land-abundant economies in which the main concern is nevertheless subsistence, the mechanization of control-intensive operations would not be expected. Even where access to markets is good, mechanization of these operations would be profitable only if wage rates were high because of nonagricultural opportunities for labor. Our field visits throughout Sub-Saharan Africa have not provided us with any instances in which control-intensive operations have been mechanized with the use of tractors. In general we found most of these operations being done by hand; in a few instances animal-drawn equipment was being used for interculture. We anticipate that in the future hand weeding may be complemented to a growing extent by animal-drawn interculture but that harvesting will continue to be done by hand.

Two Approaches to the Choice of Mechanization

In most of the literature all mechanization is seen as a choice-of-techniques issue, with the power source as the technique. Cross-sectional data are generally used to analyze the economic and social desirability of a transition from one source of power, human labor, to another, more advanced source, animal or motor power. There are two problems with this approach: the absence of consideration of farming systems in the choice of power source and the evaluation of a power source rather than the particular task for which it is to be used.

Boserup was the first to point out the importance of considering farming systems in the choice of mechanical technology. She showed, conclusively, the inappropriateness of using animal and motor power at low intensities of farming, such as forest-fallow and bush-fallow systems (Boserup 1965 and 1981). We elaborated her arguments in chapter 2 and provided evidence from Sub-Saharan Africa to show that the transition in source of power for tillage is closely associated with the evolution of farming systems. In areas of low farming intensities, therefore, the choice-of-techniques approach is inappropriate, since we are not comparing two identical farming systems, one with and one without an advanced source of power.

Consider a society that practices bush-fallow cultivation with a hand hoe. Imagine that the economic and social costs and benefits of replacing the hand hoe with animal-drawn or tractor-drawn plows for land preparation must be evaluated. In order to use the plow effectively this society would need to move to a higher intensity of farming—grass-fallow or annual cultivation. This would require substantially higher levels of land clearing and destumping and of manuring for permanent use of land. We are no longer comparing the land preparation operation with and without the plow, as the choice-of-techniques approach assumes, but rather the choice between bush-fallow and grass-fallow or annual cultivation systems.

When the technique is considered to be the power source rather than a particular task the choice between one farming system and another is being evaluated. The evolution of farming systems is determined by population densities, market infrastructure, and agroclimatic conditions and cannot be evaluated as a decision variable in the choice-of-techniques framework. The choice-of-techniques analysis is appropriate only when two power sources for doing one specific task, such as land preparation, are being compared, holding all other aspects of the farming system constant. This implies the coexistence of power sources during the transitional period and therefore the failure of attempts to replace one complete technological set, the hand hoe, with another—animal or tractor mechanization.

> Technical change cannot be viewed as the replacement of an entire set of traditional activities by a new set of modern activities. For example, bullock powered activities are rarely initially replaced in their entirety by a tractor technology even when capital is available. Instead, certain tasks are mechanized first, then others, until eventually the complete transition is made to modern farming. No aggregate concept of technology allows us to understand this microeconomic choice. What is needed is a task-by-task analysis of the profitability, availability, application and adoption of different farm practices, different machine-power combinations and different material inputs and outputs (Day and Singh 1977, 13).

Day and Singh first observed this task-specific pattern of mechanization in Punjab, India, with irrigation, land preparation, and harvesting being mechanized while cultivation and weeding were done by hand and bullock labor. I. J. Singh pursued a similar choice-of-techniques framework in Sukumaland, Tanzania, and found that a similar pattern of selective mechanization is widely prevalent in the region (Singh 1977). In table 9-2 the operations for which tractors are most often used are summarized from the South Asian literature.

When new sources of power become available, they are at first used only for selected operations for which their comparative advantage is high. Power-intensive operations, such as plowing and transport, are shifted most rapidly to new sources of power. Control-intensive operations, such as weeding, are shifted to more highly mechanized techniques when wages are high or rising rapidly (Binswanger 1984). The choice-of-techniques analysis must therefore deal with the various ways in which particular agricultural tasks can be performed and the economic costs associated with each possible way.

Chapter 10

Prospects for Tractorization in Sub-Saharan Africa

The abundance of land in Sub-Saharan Africa has often tempted developers into schemes for rapid tractorization in this continent. Many attempts at the transition to tractors in farming systems based on the hand hoe have failed because the shifting cultivation systems in which tractors were being promoted were inappropriate for use of the plow. The few instances in which introduction of the tractor was successful were areas with well-established animal-draft regimes and areas without tree stumps—valley bottoms and grassy savannas.

Recall from chapter 2 the reasons for the inappropriateness of using the plow in the long-fallow systems of cultivation—the forest-fallow and bush-fallow systems. For use of the plow to be profitable it would be necessary to practice intensive permanent cultivation in areas where shifting cultivation is the norm. Much overhead labor would be required for the destumping of fields and for maintenance and training of animals. Permanent cultivation of land would also require high levels of labor for weeding and for maintaining the fertility of the soil. The transition from the hand hoe to the plow, whether animal-drawn or tractor-drawn, therefore, is not cost-effective in areas under long-fallow systems of cultivation.

As fallow periods decline because of population growth or improved access to markets, the costs of adopting a plow-based farming system decline on account of a reduction in destumping costs, an increase in the availability of fodder, and the introduction of soil-fertility restoration techniques. Also, as fallow periods decline the benefits of adopting the plow increase because of the labor savings associated with land preparation and weeding. The third question to be taken up in this chapter is whether it is possible to go directly to tractors in farming systems in which use of the plow is becoming appropriate or whether the transition to an animal-traction system is more cost-effective and therefore more likely to succeed.

Replacing Animal Draft with Tractors: The Choice of Techniques

Unlike the transition from the hand hoe to animal power the shift from animal power to tractors has been studied quite thoroughly; see Binswanger (1978) and Herdt (1983) for recent reviews of these studies for South and Southeast Asia and Binswanger (1984) for a worldwide historical review. Although this transition has not been the focus of much attention in Sub-Saharan Africa the primary issues are likely to be the same as in Asia and elsewhere. This is because the farming systems for which tractors are appropriate are likely to be the same, as are the analytical techniques used for studying this transition.

Consider a society with a well-established farming system based on the animal-drawn plow. We have already seen that this society is characterized by intensive production of field crops on permanently cultivated plots of land. We also know that animal-draft power is used selectively for power-intensive operations such as tillage and transport, while other operations continue to be done by hand. When tractors become available in this society they will be used primarily as a substitute for animal draft and for labor in these power-intensive operations, while all other agricultural operations will continue as before.

We are now in the realm of the familiar choice-of-techniques problem, where the costs of alternative sources of power for performing specific tasks is being compared rather than the total costs of production using alternative farming systems. The economic costs of using tractors instead of animals is determined by the following variables: the relative costs of labor and capital, the interest rate, the utilization of capacity, farm size, the availability of fodder, the comparative maintenance costs of animals and tractors, and the difficulty of obtaining spare parts, fuel, and repair services.

Let us illustrate the choice-of-techniques problem by examining the choices made by a farmer in Sukumaland, Tanzania. We rely on the analysis of I. J. Singh (1977) for much of the data required for this example. Let us suppose that this farmer has five hectares of land that he cultivates annually without letting it lie fallow. Suppose also that the area has an adequate supply of trained oxen for purchase and private tractor-rental services.

Preparatory tillage is a prerequisite for seeding on this land, since it is cultivated permanently. The choices facing the farmer are to till his soil manually, using a hand hoe; to till his soil with an ox-drawn plow; and to hire a tractor. Assume that the differences in yield and timeliness among the three arrangements is minimal. The dominant possibility will therefore be the one that offers the lowest cost per hectare.

Table 10-1 provides a cost comparison among the three sources of power at respective opportunity costs of labor of 5, 10, and 15 Tanzanian shillings a day. At an opportunity cost of labor of 5 shillings, the ox-drawn plow is marginally superior to the hand hoe, and a tractor-rental arrangement is decidedly inferior. Manual techniques will supersede animal draft power if the labor supply increases, thereby lowering the opportunity cost of labor further, if the cost of animal maintenance rises because of a scarcity of fodder or high incidence of animal disease, or if the price of plows rises.

At an opportunity cost of 10 Tanzanian shillings a day ox-drawn plows are definitely superior to manual techniques. Their superiority to tractor-rental arrangements, however, depends on the total number of hours that the tractors work. In areas in which the utilization rates are high the cost of operating and amortizing the tractor is shared by a

Table 10-1. *The Choice of Sources of Power for Plowing a Five-Hectare Farm, Sukumaland, Tanzania*

		Cost per hectare		
Source of power	*Work time required per hectare*	*Opportunity cost of labor 5 shillings a day*	*Opportunity cost of labor 10 shillings a day*	*Opportunity cost of labor 15 shillings a day*
Manual tillage				
Labor	66 workdays	330	660	990
Ox-drawn plow				
Labor	40 workdays	200	400	600
Oxen	7.5 ox-pair days	124	124	124
Total		324	524	724
Tractor rental A[a]				
Labor	18.2 workdays	91	182	273
Oxen	5.5 hours	427	427	427
Total		518	609	700
Tractor rental B[b]				
Labor	18.2 workdays	91	182	273
Oxen	5.5 hours	354	354	354
Total		445	536	627

a. The tractor is assumed to work 1,000 hours a year.
b. The tractor is assumed to work 1,500 hours a year.
Source: The primary data were drawn from Singh (1977).

large number of users and the unit costs of contract hire are lower. In rental arrangement A, for example, the tractor works 1,000 hours a year; at this utilization rate animal draft dominates tractor rental, but when tractor use goes up to 1,500 hours a year—rental arrangement B—animal draft is only marginally superior.

If the opportunity cost of labor goes up to 15 Tanzanian shillings a day, a tractor-rental arrangement is definitely superior to the ox-drawn plow. The higher the rate of utilization of the tractor, the wider the gap between the two techniques becomes. There are two factors, however, that could limit the advantage of the tractor over the ox-drawn plow—the manuring benefits to be derived from the use of oxen and the risk of not obtaining the tractor on time.

The cost of manure to farmers who use animal draft is lower, so they may continue to use the animal-drawn plow, even if to rent a tractor for tillage might be slightly cheaper. Animal draft may continue to be superior when the risk of not obtaining a tractor on time is monetized in the cost equation. We have already seen that rental markets are difficult to establish for time-bound and synchronic operations, such as plowing, because of sharp conflicts among potential users over timing. This conflict arises because it is impossible to decide in advance the best sequence in which to serve different customers of the same rental service. Farmers who rely exclusively on a rental service face the risk of delayed sowing and the consequent reduction in yields because the tractors did not reach them in time.

Let us now turn to the next choice, which is among using one or more pairs of oxen, renting a tractor, and owning a tractor. Given the lumpy nature of the purchase of a tractor, the profitability of ownership is determined by the size of the machine, the size of the farm, and utilization of the capacity of the equipment.

Consider first an environment in which rental markets are difficult to establish—because of a short plowing season, for example—and where farms are small. Tractors, to be effective in this environment, must be tailored to the size of the farm. In other words, the feasibility of using small four-wheel tractors of 20–25 horsepower and power tillers must be considered. Worldwide experience has shown that tractors of less than 35 horsepower and power tillers are not suitable for upland field-crop production.[1] Animal-drawn plows will therefore continue to be the only workable alternative to manual tillage in such environments.

The second environment to be considered is one in which rental markets are easily established and farms continue to be small. If tractor tillage is dominant over animal draft in this environment, the farmer must decide whether to buy or merely to rent a tractor. This decision must be based on the rate of return on an investment in a tractor in relation to the refund on other possible investments, the cost of maintenance of the tractor, and the cost of organizing and supervising rental of the tractor to other farmers.

Where farms are larger, ownership of tractors becomes dominant, even in environments with lower utilization of capacity. If possible, the size of the tractor should be based on the expected capacity utilization. The emergence of 35-horsepower tractors in the Indian Punjab, for example, was in direct response to the time-bound nature of the plowing operation and the sizes of the farms, which are smaller than U.S. farms.

In closing this section, let us reiterate that there is no single answer to the question of when it is appropriate to move from animal draft to the tractor. The economics of such a move ought to be analyzed case by case. It must again be emphasized, moreover, that when wages are low, the transition from animal draft to the tractor will not lead to a complete replacement of animal draft, because farmers will continue to exploit a complementarity between animal draft and the tractor that arises out of the specialization of the tractor for power-intensive tillage and for transport.

Effects of the Transition from Animal Draft to the Tractor

In chapter 7 we discussed the farm level benefits of a transition to the animal-drawn plow. Discussion of the farm-level benefits of a transition from animal draft to the tractor can be similarly organized around table 7-2: private and aggregate area-expansion benefits, the potential yield effect, and the labor-savings effect.

The Private and Aggregate Area-Expansion Effect

Recall briefly the difference between private and aggregate area expansion: expansion of the aggregate area occurs only when private farms extend into fallow or reclaimed land or when the cropping intensity on a given plot of land is increased. When private area expands at the cost of other farmers there are no aggregate area-expansion benefits (see table 7-2). Any inferences concerning the aggregate expansion in agricultural output through an increase in land area cultivated made possible with the use of the tractor would require additional information on the source of the additional area. McInerney and Donaldson (1975), for instance, found that in Pakistan expansion in the area cultivated by tractor-using farmers came about mainly through the transfer of control of already farmed land, and Jabbar and others (1983) reported that 81 percent of power tiller owners in Mymensingh, Bangladesh, increased their cultivated area by expanding into land previously rented out. We are not aware of any studies conducted in Sub-Saharan Africa in which it was determined where the additional area came from.

In areas in which few opportunities exist for expansion of farms, the effect of additional power on cropping intensity is often regarded as an important potential benefit, achievable mainly through faster cultivation between seasons. Accordingly, in several studies conducted in land-scarce economies the attempt has been made to determine the effect of the transition to tractors on cropping intensity (Farrington 1985, Herdt 1983, and Binswanger 1978 review such studies conducted in South and Southeast Asia). In South Asia, of the sixty-three observations in which the effect of tractors on cropping intensity is documented, 73 percent fall into the "no clear difference" class of minus 10 percent to plus 10 percent (Binswanger 1978, 11). Herdt's review of Southeast Asian cases confirms this finding: nine of the ten instances that he reported show no significant increase in cropping intensity.

Even in instances in which significant increases in cropping intensity were observed the effect of irrigation cannot be isolated. In the Surat district in India, for example, tractor owners show a statistically significant intensity difference of 13.9 percent over bullock owners, but that is associated with a large difference in their use of irrigation—22 percent of the gross cropped area rather than 60 percent (Binswanger 1978, 25). It should also be noted that extremely high intensities of cultivation have been achieved without the use of the tractor—in South Sulawesi, for instance, cropping intensities were around 1.92 before and after the introduction of the tractor (Herdt 1983). It would be difficult, therefore, to attribute the increases in cropping intensity solely to the transition to the tractor.

In most land-scarce areas the constraint on higher cropping intensities is not the lack of power—to achieve a faster turnaround between crops—but rather the lack of moisture in the absence of irrigation systems. Substantial gains in intensity for rainfed systems could therefore not be expected in the semiarid zone. In the humid tropical lowlands and the tropical highland areas, however, crops can be grown the year round, and high cropping intensities have been achieved without tractors.

The only area in which tractors could have an effect on intensity is in the bimodal rainfall areas, where rainfall precedes each crop; there are winter rains and summer rains. The summer crop needs to be harvested before the winter rains begin in order to minimize the risk of grain spoilage and to clear the fields of the crop before the land is prepared for the winter crop. In such areas, a combination of the tractor and the power thresher could relieve the labor bottleneck during the harvesting and land-preparation period. High intensity of cropping has been sustained in Thailand and the Philippines by the use of motorized sources of power for land preparation and threshing (Juarez and Pathnopas 1983). Irrigation and high-yielding varieties made double cropping possible. With the first crop usually harvested during the rainy months, the danger of grain spoilage was high. Also, fields should be cleared of the previous crop before the irrigation water is released.

Yield Effect

As in the transition from the hand hoe to the animal-drawn plow, the yield increases associated with use of the tractor could come about if the quality of tillage were improved. Quality of tillage can be improved if a switch from the animal-drawn to the tractor-drawn plow were made for one or more of the following reasons: on a plot of land, where deep plowing affects the yield, shallow plowing is replaced by deep plowing; a change from a disc plow to a moldboard plow, for example, could improve yields because the moldboard incorporates crop residues better, and a more timely completion of land preparation leads to higher yields.

The existing evidence indicates, however, that there is generally no significant difference in yield on a given farm whether animal draft or the tractor is used. Herdt (1983) found that the difference in yield on farms that used animal draft and those on which the tractor was used was negligible after differences in fertilizer use had been taken into account (see table 10-2). This is consistent with the results from South Asia, where of 107 comparisons, more than 50 percent showed significantly higher yields; almost all these instances, however, were associated with an equal or greater percentage increase in fertilizer use (Binswanger 1978, 11). Recall that on certain soils and crops, improvements in the quality of tillage do improve yields, as has been shown in many experiments. If differences in yields are not observed between farms on which animal traction is used and those on which the tractor is used it must be concluded that the transition to the tractor-drawn plow is rarely motivated by the desire to improve the quality of tillage. Area expansion or labor saving must be the driving forces behind such a transition.

Table 10-2. *Summary of Studies in Which Rice Yields on Farms on Which Land is Prepared with Animal Traction or by Hand and Those on Which Mechanical Methods Are Used Are Compared*

Region or country	*Comparison*	*Reported yield (tons per hectare)*	*Fertilizer*[a] *(kilograms per hectare)*	*Adjusted yield (tons per hectare)*
Nepal	Bullocks	1.7	16	1.7
(without pumps)	Tractors	2.1	164	1.4
Nepal	Bullocks	2.1	214	2.1
(with pumps)	Tractors	2.3	264	2.1
West Java, Indonesia	Animal	4.9	323	4.9
(wet, 1979–80)	Tractors	4.9	323	4.9
West Java, Indonesia	Manual	3.8	285	3.8
(3 seasons, 1979–80)	Tractors	3.9	308	3.8
South Sulawesi, Indonesia	Animal	2.7	138	2.7
(3 seasons, 1979–80)	Tractors	2.9	227	2.5
Nueva Ecija, Philippines	Carabao	2.6	89	2.6
(wet, 1979)	Tractors	4.0	129	3.8
Central Thailand	Bullock	2.6	32	2.6
(irrigated, transplanted)	Tractors	2.8	48	2.6
Central Thailand	Bullock	0.2	3	0.2
(rainfed, broadcast)	Tractors	0.2	2	0.2
Bangladesh	Bullock	1.5	n.a.	1.5
	Power tiller	1.5	n.a.	1.5
Central Luzon, Philippines	Before tractors and tillers	2.2	57	2.2
	After tractors and tillers	2.6	79	2.1
Eight provinces, Philippines	Animals	2.6	86	2.6
	Tractors and tillers	2.8	117	2.5

n.a. Not available.
a. Urea.
Source: Herdt (1983).

Changes in Labor Use

From the literature we compiled twenty-four observations on labor use, by operation, on farms on which animal draft and the tractor were used in order to ascertain the changes in labor use, by operation and by total labor use, and shifts in the amount of labor used between operations. Does a reduction in the amount of labor required for land preparation, for instance, lead to an increase in the amount of labor required for weeding? Summary information is given in table 10-3.

Table 10-3. *The Change in Labor Use Brought About by a Shift from Animal Traction to the Tractor*
(number of instances)

	Percentage change							
Agricultural operation	< −75	*−75 to −50*	*−50 to −25*	*−25 to 0*	*0 to 25*	*25 to 50*	*50 or more*	*Total*
Land preparation	9	4	11	0	0	0	0	24
Planting, sowing, transplanting	4	3	1	11	3	0	0	22
Irrigating, fertilizing, applying pesticides	0	0	11	6	0	1	1	19
Weeding and interculture	1	0	4	12	6	1	0	24
Harvesting, threshing, transport	0	0	3	18	2	1	0	24
Other labor	0	0	1	1	0	2	2	6
Total labor	0	2	9	11	2	0	0	24

In twenty-two of the twenty-four studies considered labor use per hectare of crop production was reported to be lower on farms where the tractor was used than on those on which animal draft was used. In twelve studies reductions in labor use of 50 percent or more were reported. This is consistent with the empirical findings presented in chapter 7 (see table 7-4), where by means of a worldwide data set we showed that a transition to the tractor is associated with significant reductions in labor use per hectare.

The greatest reduction in labor use was in land preparation. For all studies in which reductions in labor input of at least 25 percent were reported, in nine, reductions that exceeded 75 percent were reported. Consider instances in which reductions in labor use of 50 percent or more for land preparation were reported and trace through the effect on other operations (see table 10-4). Consider weeding first: of the fourteen instances in which land-preparation labor was reduced 50 percent or more, in only two was there a reduction in weeding labor greater than 25 percent. In ten instances reductions smaller than 25 percent in weeding labor and in two instances increases in weeding labor over that required with animal traction were reported. The situation is similar with respect to harvesting: of the fourteen instances in which labor for land preparation was reduced 50 percent or more, in only one was there an equal reduction in harvesting labor. In nine instances reductions in labor of less than 25 percent were reported, and in one labor requirements were found to have increased.

These findings indicate that the labor savings brought about by the transition to the tractor are concentrated primarily on land preparation, where a substantial reduction in labor peak is observed. Where an aggregate area-expansion effect exists, however, a rightward shift in the demand curve for weeding and harvesting labor could be expected, despite the lower requirements per hectare.

Income and Income-Distribution Effects

The income effects and income-distribution effects of the transition to the tractor depend on the elasticity of demand for output, transport infrastructure, and the distribution of factor ownership. Table 10-5 lists the income effects of a transition from animal power to the tractor in land-scarce economies. For the sake of simplicity we shall assume that rental markets for tractors do not exist.

Consider first the farmers who face a downward-sloping demand curve for their output. The adoption of technologies that bring about an expansion in food output depresses its price so much that the incomes

Table 10-4. *The Change in Labor Use Brought About by a Shift from Animal Traction to the Tractor in Instances in Which the Change in Land Preparation Labor Is Less than –50 Percent*
(number of instances)

Agricultural operation	*Percentage change*						*Total*
	< –75	*–75 to –50*	*–50 to –25*	*–25 to 0*	*0 to 25*	*25 to 50*	
Land preparation	9	4	0	0	0	0	13
Planting, sowing, transplanting	4	3	1	4	0	0	12
Irrigating, fertilizing, applying pesticides	0	0	9	0	0	1	10
Weeding and interculture	0	0	2	9	1	1	13
Harvesting, threshing, transport	0	0	3	9	0	1	13
Other labor	0	0	0	0	0	1	1
Total labor	0	2	9	1	1	0	13

Table 10-5. *Land-Scarce Economies: The Effects of Mechanical Technology on the Distribution of Income*
(direction of effect)

Expected effect	*Landless labor*	*Landowners*	*Tenants*	*Nonfarm consumers*
Inelastic demand for food				
Expansion of output	+	–	–	+
Labor savings only—no expansion in output	(–)	+	–	(+)
Subsidies for machines and equipment	(–)	+	–	(+)
Guaranteed support price	(+)	+	–	–
Elastic demand for food				
Expansion of output	+	+	–	?
Labor savings only—no expansion in output	(–)	+	–	?
Subsidies for machines and equipment	?	+	–	?

Note: Signs within parentheses indicate the direction the effect is expected to take in most instances, though some uncertainty persists. A question mark indicates that the direction of the effect is uncertain.

of producers are reduced while the real incomes of net buyers of food in rural areas, such as landless laborers and urban consumers, are increased. If the result of the transition to the tractor is a saving in labor rather than an expansion in output, however, large landowners will gain at the expense of laborers and tenant farmers.

The situation is different where the demand for food is perfectly elastic; the net beneficiaries in this instance are the tractor farmers. The effects on labor depend on whether the benefits of an expansion in aggregate area are more or less in line with the benefits of the savings in labor. In areas where area expansion is more important, the demand for weeding and harvesting increases, so the net income of labor increases.

Transport infrastructure is as important as elasticity of demand in determining the profitability and the consequences for distribution of a transition to the tractor. In areas in which transport facilities are extremely poor it would not be possible to capture the area-expansion benefits to be realized with the tractor and the demand curve would therefore be inelastic, even if the regional or national demand curve were perfectly elastic. The transition to the tractor leads to gains in income in areas in which transport facilities are good, at the expense of poorly connected areas.

The net beneficiaries of the switch to the tractor in most instances are the large landowners. These farmers have easier access to credit and information and can capture the economies of scale inherent in a lumpy input such as a tractor. Tenant farmers are usually subject to negative income effects because these farmers are displaced by the landowers. The consequences of tractor subsidies and favorable credit terms for the distribution of income are therefore highly negative. This would also be true if support prices for agricultural output were higher than market prices.

Table 10-6 presents the income effects of a transition to the tractor from the animal-drawn plow in land-abundant economies. The situation differs here mainly in the factors that are considered scarce and in inelastic supply. Land is no longer in inelastic supply, so large landowners do not necessarily accrue the rents from tractorization. The main beneficiaries are wealthy farmers or those who have access to credit. In Sub-Saharan Africa these are usually owners of large herds of cattle. Since there is no landless labor class in a land-abundant economy displacement of labor as a result of mechanization is not observed (Binswanger and McIntire 1986). The demand for seasonal labor for

Table 10-6. *Land-Abundant Economies: The Effects of Manual Technology on the Distribution of Income*
(direction of effect)

Expected effect	*Seasonal labor*	*Farmers having large herds of cattle*	*Farmers having small herds or no cattle*	*Consumers off the farm*
Inelastic demand for food				
Areas in which transport is poor				
Expansion of output	(+)	–	–	(+)
Labor savings only—no expansion of output	(–)	–	–	(=)
Areas in which transport is good				
Expansion of output	+	(+)	–	+
Labor savings only	(–)	+	–	(+)
Existence of rental markets	–	+	+	(+)
Elastic demand for food				
Expansion of output (no rental markets)	+	+	–	?
Labor savings (no rental markets)	–	+	–	?
Existence of rental markets	–	+	+	?

Note: Signs within parentheses indicate the direction the effect is expected to take in most instances, though some uncertainty persists. A question mark indicates that the direction of the effect is uncertain.

weeding and harvesting usually increases with use of the tractor because larger areas are cultivated; the income effects for labor could therefore be positive. The situation would be the same as that of nonfarm consumers in land-scarce economies.

There are no tenant farmers in a land-abundant economy, where access to land is not restricted; cattle holdings, however, could be skewed, so many farmers do not have sufficient wealth or access to credit to make the transition to the tractor. These farmers cannot capture the area-expansion and labor-saving benefits of the tractor and many suffer negative income effects. In areas in which there are rental markets these farmers can increase their income to some extent by using tractor services on contract hire.

Tractorization at the Late Bush-Fallow and Early Grass-Fallow Stages

We argued earlier that the most appropriate stage in the evolution of farming systems to introduce the plow is at the late bush-fallow or early grass-fallow stage. The question here is whether animal-draft power can be bypassed at this stage in favor of a direct move to the tractor. The conclusion is that the direct transition to the tractor is not cost-effective in most environments. The exceptions are the treeless flood plains and grassy savannas.

Recall our farmer who is faced with a choice of using manual, animal-draft, or tractor power for the tillage operation. Consider first a choice between using animal draft and renting a tractor. Then, in addition to the choice-of-techniques issues discussed in the preceding section this farmer must evaluate the costs of destumping for animal and tractor tillage, the costs of animal maintenance, and the cost of losing the complementarity between tractor and animal.

Let us first deal with destumping costs. These are higher for tractor farmers than for those using animal draft, because a higher quality of destumping is required in order to minimize the damage to tractor-drawn implements. Animal-draft farmers can work around the stumps and lift their plows over them, so animal-drawn plows can be used even in short-fallow systems. In time, as the fallow periods on these fields decline, the stumps decay and the fields become stump-free. During our field visits, we tried to ascertain whether the presence of a few tree stumps was a constraint to the adoption of the plow. In almost all instances in which the question was relevant stumps were not a problem and did not prevent the adoption of the plow.

Since the destumping costs associated with animal traction are small, periodic fallowing after eight or ten years of cultivation is still feasible. In areas in which use of the plow is emerging, moreover, field conditions are varied. Some fields are permanently cultivated, while bush fields are still periodically fallowed. The use of animal-drawn plows allows the farmer to till both his permanently cultivated fields and his bush fields.

Destumping costs vary with rainfall and population density; they are associated positively with the former and negatively with the latter. Table 10-7 presents examples of destumping costs from selected locations in Sub-Saharan Africa. They are very high in the sparsely populated humid tropics—$600 a hectare in the Côte d'Ivoire and $850 a hectare in Guinea—while they are extremely low in the semiarid regions—$25 a hectare in Central Sudan—where rainfall is low. Destumping costs are lower in densely populated areas because of shorter fallow periods (see the examples from Nigeria in the table). Destumping costs as a constraint to the direct transition to the tractor in the late bush-fallow and early grass-fallow stages is more pronounced in the more humid zones and less pronounced in the more densely populated areas.

By the late bush-fallow and early grass-fallow stages, the supply of animals to the area increases substantially because of the ease of obtaining fodder and a reduction in the incidence of trypanosomiasis. Animal-maintenance costs are therefore considerably lower than in the long-fallow systems and the farmer can benefit from various types of interaction, such as production of manure, between crops and livestock. It is therefore likely that the farmer has already begun to keep animals on his farm. By choosing not to use animals for draft the farmer would forgo an additional, important benefit that he could derive from incorporating animals into his farming system. This argument would, of course, be less strong in areas in which the incidence of animal diseases, such as rinderpest and black lung, is high.

By going directly to the tractor the farmer would also lose the complementarity between tractors and animals. Recall that during the early stages of tractorization, tractors are used for power-intensive operations, while animals continue to be used for interculture and weeding. This complementarity is particularly useful in areas where wages are low.

Now consider a farmer who is trying to choose between using draft animals and owning a tractor. In addition to all the foregoing he must consider two further issues: the learning costs associated with animal

Table 10-7. *The Costs of Clearing Land in Selected African Locations*

Country and region	*Annual rainfall (millimeters)*	*Rural population density (persons per square kilometer)*	*Costs per hectare (1980 U.S. dollars)*
Côte d'Ivoire			
South	1,500	12.0	600
Guinea	2,200	11.4	850
Sudan			
Gedaref	400–800	n.a.	25
Kenya			
Narok	1,000	15.0	400
Nigeria			
Kwara	1,200	13.0	350
Ayonba	1,200	76.0	100
Lafia	1,200	53.0	180

n.a. Not available.
Source: World Bank data.

draft and use of the tractor and the existence of infrastructure for efficient tractor operations. In general, the learning costs associated with use of the tractor are higher than the cost of learning to use animal draft in the farming system. The operation and maintenance of tractor equipment is certainly more complicated than the use and maintenance of animals and animal-drawn equipment. The chance of making a mistake that will cause costly damage to the equipment or lead to a loss in output is substantially greater with tractors than with animal-draft equipment.

For the efficient transition from hand power to tractor power it is essential to have high levels of investment in such infrastructure as on-farm and off-farm roads, fuel supply, and repair services. These infrastructure costs are typically higher in areas in which population density is low. The level of infrastructure required is much lower for animal draft; as population density increases, investments in this kind of infrastructure tend to be made, so in time the transition to the tractor becomes more cost-effective.

Direct Transition to the Tractor in a Treeless Environment

There are two areas in which the introduction of the plow is not constrained by high destumping costs, even when population density is low: valley bottomlands in which irrigated or flooded rice is cultivated and the grassy savannas in the semiarid zone. The natural vegetation in these areas is primarily a grass cover, not including bushes or trees. Such land can be put under the plow directly, without the enormous overhead expenditure of destumping. Further case-by-case analysis is required, however, to determine whether the animal-drawn plow or the tractor is appropriate in a given location. The choice between the two is based on the choice-of-techniques question, the dynamic learning costs, and the cost of infrastructure.

Irrigated or Flooded Rice

As mentioned earlier, the deep and heavy soils of the valley bottoms or flood plains are highly responsive to intensification inputs, and the yield benefits of plowing are high, especially for rice. It should be noted, however, that these lands may require substantial investments in drainage, water control, and irrigation before they can be brought under cultivation. The rate of return on these investments is highest where population density is high or high market demand for food is sustained. Examples of such valley bottoms or flood plains that are now under mechanized rice cultivation are Hauts Bassin, Burkina Faso; Casamance, Senegal; Odienné, Côte d'Ivoire; Mopti, Mali; Lake Victoria Basin, Tanzania; and northern Ghana.

Consider the case of northern Ghana. The area within a seventy-mile radius of Tamale is a flat alluvial plain crisscrossed by several streams. This land suffers from impeded drainage and is flooded during the rainy season. Tall grasses and very few trees grow there, and until the early 1960s the plain was used exclusively for dry-season grazing. Since then, the area has been rapidly brought under rice production following an intensive extension effort by the government and the provision of loans for tractors at a subsidized interest rate. Today this entire area is under paddy rice production. Tractors are owned by several large farmers and each machine is used to prepare an average of 300 acres a season—both the area owned by the tractor owner and the area that he plows through contract-hire operations. For further details see WARDA (1977).

Animal draft is used to a moderate extent in northeastern and northern Ghana (Munzinger 1982). Why, then, have tractors been found more profitable than animals in this area? The reasons are that farms are larger on the plains than on the uplands; contract-hire operations are possible because the greater water-retention capacity of the clayey soils lengthens the growing season; capital costs have been reduced by subsidized interest rates; and the costs of infrastructure have been reduced by government-provided repair, spare parts, and input services and by a government marketing board for the output.

Unlike northern Ghana where cultivation of the bottomlands exclusively is the norm, in most instances bottomlands are cultivated in combination with upland farming. In Sierra Leone, for instance, small plots of rice are cultivated on the bottomlands using animal draft, while the upland areas continue to be cultivated with the hand hoe using fallow techniques. Animal draft is cost-effective because the plots are small and because its use can be introduced on the upland farms as they become permanently cultivated. Similar developments can be anticipated in the swampy valley bottoms in the humid zones of Nigeria, Guinea, Côte d'Ivoire, and Liberia.

In areas where valley bottom plots are relatively small in growth the use of the power tiller as a substitute for animal-draft power can be anticipated as incomes rise. The use of such equipment for mechanical puddling of rice fields has become common in parts of Southeast Asia. In Thailand, for instance, more than 100,000 power tillers are now being used by small farmers (Smeyers 1979). Since the 1960s small rice farms in the Philippines have similarly shifted to the use of power tillers (Barker and others 1973). Similar possibilities exist for the valley bottomlands of Africa in the long term.

Mechanized Rainfed Farming in Sudan

The grassy savannas of Sub-Saharan Africa are another area in which the potential for mechanized crop cultivation was initially thought to be extremely high. Rainfed mechanized farming in central Sudan was the only instance in which this expectation was fulfilled. The central clay plains of Sudan offer a unique example of direct transition to the tractor without any intermediate animal-power stage (Tothill 1948, Agabawi 1968, Mohammed 1982). Approximately 3 million hectares in the Central Plains—Kordofan, western Sudan—are under mechanized farming, mainly by individual farmers. In 1982 this rainfed mechanized sector produced more than 70 percent of Sudan's sorghum and 30 percent of its sesame.

The central plains are in the semiarid zone, with annual rainfall of 500–800 millimeters and growing periods of between 92 and 120 days. Two main soil types are found there, the *Qoz* sands, which are light and well drained, and the heavy clays, which are hard when dry, plastic when wet, and extremely difficult to work. Before the introduction of tractors the clay plains were used mainly for grazing, while farming was concentrated on the light sandy soils. The sandy soils were preferred, despite their lesser fertility and greater risk of drought, because of the ease and speed with which they could be worked.

A mechanization scheme was introduced in this area in 1944, and for a nominal fee individual farmers were allowed to lease blocks of land, 420 hectares each, for mechanized farming of sorghum. The acquisition and use of tractors was promoted through Mechanized Crop Production Schemes (MCPS). Mechanized production was made highly profitable by a government-supported scheme for the transport of grain from the farm and export to Saudi Arabia, where Sudanese sorghum commands a premium price, and to other countries. Exports in 1982 were around 20 percent of total production. The present shadow exchange rate is about US$1 to 2 £Sd, and it is estimated that even at an average yield of 762 kilograms a hectare the domestic resource cost of mechanized production of sorghum would be less than $1 in the Jeddah market and equal to $1 at the world market price. The situation would be even more favorable for yield levels of 1,000 kilograms a hectare or more.

To summarize, there are three primary reasons for the rapid growth of the area under mechanized farming: before 1944 the natural vegetation of the area was a grassy savanna with trees, mainly acacias, sparse, so the cost of destumping was not a constraint to expansion of the area under cultivation; to produce even a modest yield of sorghum on these soils tillage is required, but the soils are very difficult to cultivate in the absence of an advanced source of power; and because of the greater water-retention capacity of these soils the number of days available for plowing is somewhat greater than on sandy soils. The farmers could therefore achieve high rates of utilization of their equipment and spread equipment costs over a larger cultivated area. To these was added the fact that the several million hectares of grasslands involved were not claimed by any of the generations of families living in the area.

Could an animal-draft system have succeeded here? It could have if the domestic resource costs of using animal draft had been lower than the domestic resource costs of using tractors. We do not know of any analysis in which this comparison was made. Northern Sudan, especially in the area next to the Egyptian border, is accustomed to using animal-draft power; the technique was not new to the country. On the central plains, moreover, there were large herds of cattle because of the ease of obtaining fodder, so the supply of oxen was not a constraint. Labor supply was highly constrained, however, and the benefits of area expansion were great, so tractors were the predominant technology. Whether the mechanized crop-production scheme will work for Sudan in the long run will depend on whether it will encourage the intensification of existing fields rather than abandonment of them in favor of newly cleared fields and whether the spread of striga can be controlled through the application of appropriate pesticides.

Note

1. Power tillers and small four-wheel tractors have been used successfully in cultivation of flooded and irrigated rice, as seen in South and Southeast Asia; in small-scale production of vegetables—common in Europe; for preparation of the soil in the hills of Switzerland; and for transport, for which they are becoming increasingly common for the delivery of goods in urban areas of India. For a detailed analysis of the economics of the use of the power tiller, see National Council of Applied Economic Research (1977).

Chapter 11

Mechanization Choices for the Humid Tropics

Most soils in the humid tropics are not suited to intensive production of field crops and are therefore inappropriate for use of the plow. What, then, are the choices of mechanization for the humid tropics?

The constraints to intensive production of field crops in the humid tropics have already been mentioned. Most soils of the subhumid and humid tropical lowlands have low cation-exchange capacity (CEC), and continuous cultivation of field crops leads to rapid leaching, acidification, or erosion, causing fertility, and therefore yields, to decline (Allan 1965; National Research Council 1982; Ruthenberg 1980; FAO 1984; Kang and Juo 1981).

Many attempts have been made to establish permanent production of field crops on these soils with mechanized clearing and land preparation, but after a few years problems have generally been encountered. Early experiments of this kind were carried out by the Belgian Colonial Administration in Yangambi, Zaïre, in 1960. One hundred fifty acres of forest was cut down, burned, and uprooted. The soil was deep-plowed and cultivated in a manner similar to the method followed in the temperate zone. Rice was planted, followed by manioc, then groundnuts. By the second year the yield of every crop was found to have dropped substantially. Deep plowing had impaired the structure of the soil, and the exposed soil had been severely leached. The entire 150 acres were abandoned after a few years, and ten years later the area was only thinly wooded (Ruthenberg 1980, 63). Similar projects for initiating continuous, mechanized cultivation were quickly abandoned in the humid zones of Côte d'Ivoire, Nigeria, and Ghana (see FAO 1984 for the first two and Moorman and Greenland 1980 for Ghana). In all cases the projects produced extremely low returns on investment and very high levels of environmental degradation.

Note that permanent production of field crops on most of these soils whose CEC is low is technically possible, and a balanced system can be achieved with heavy use of fertilizers, pesticides, and other purchased inputs. But the total costs of such a high-input system per unit of output are typically higher than under shifting cultivation. Volcanic soils, alluvial soils, and vertisols, however, have high CEC and can sustain intensive production of field crops with low levels of degradation.

What are the economically sound choices for increased food production in the humid tropics in the face of increasing population density? In general, a system of farming that closely mimics the natural vegetation is what will work for the long run. A system of intercropping of short-term and medium-term crops—such as manioc, yams, maize, rice, pigeon peas, and chick peas—under a protective cover of trees, for example, is feasible, since the ground would be covered for most of the year and detrimental effects on the soil would thus be minimized (Okigbo 1974). Technical choices for permanent cultivation on the low-CEC soils of the humid tropics therefore differ for areas in which markets for tree crops exist from choices for areas in which they do not.

Technological Choices with Tree Crops

The humid and subhumid tropics are well suited to perennial crops such as bananas and to tree crops such as rubber, cocoa, and oil palm. Once these crops have been established their canopy reduces the growth of weeds and they demand less of the soil than annual crops. The effects on the soil resemble those of forest and bush vegetation in establishing a closed nutrient cycle (Ruthenberg 1980, Sanchez 1976). The tree cover, moreover, protects the soil from erosion. Food crops can be intercropped during the establishment phase of a tree crop and sometimes on the mature plantation. In southeast Côte d'Ivoire much of the marketed plantain is produced by intercropping young cocoa plantations. Once the cocoa trees have been established production of plantain is moved to other newly planted cocoa plots. Similarly, Yoruba farmers in Nigeria grow yam, manioc, and maize between cocoa plants during the first five years (Ruthenberg 1980). Ruthenberg lists the following four phases in the establishment of smallholder tree-crop plantations:

1. Perennial crops are planted around the house as the shifting cultivator begins to maintain a permanent residence.
2. The area devoted to perennial crops is extended by interplanting perennials with annuals in newly cleared plots.

3. In time the perennial crops predominate in the mixed cropping system, and young stands of tree crops are interplanted with arable crops.
4. As perennial crops create more and more shade, intercropping becomes less important.

The relevant question here is, What operations can be mechanized in a farming system in which tree crops are dominant? Transport and food-processing operations are the most power-intensive in this system and are the ones most likely to be transferred to motorized power. Such has indeed been the case in the humid tropics of West Africa, where roads were built into cocoa- and oil-palm–producing areas and the products are transported by motor truck. During our field visits we observed wide use of mechanical mills in Côte d'Ivoire, southern Nigeria, and Guinea. In a farming system dominated by tree crops land preparation is not a major operation, for it is restricted to areas between young trees. In smallholdings this operation can be accomplished with the hand hoe because of the softness of the soil and the length of the period available for land preparation.

In discussions of the importance of tree crops as a choice for the humid tropics it should be borne in mind that the returns on tree crops are determined by farmgate prices. The establishment of tree crops is not feasible where well-developed infrastructure for exporting the crop is not available. Tree crops are also not a wise choice in the face of falling or widely fluctuating export prices. High taxes on rubber exports, for instance, were instrumental in the abandonment by many smallholders of rubber gardens in the humid zone of Nigeria (Moorman and Greenland 1980).

Technological Choices without Tree Crops

Agrosilviculture is a regulated system of plantation forestry, in which a public authority allocates forest land to shifting cultivators to clear. The cultivators are allowed to plant for a limited time, after which the cultivators are required to plant tree seedlings and to tend them until the trees are large enough to choke the regrowth of weeds. The system known as Taungya is essentially a form of regulated shifting cultivation, in which the farmer is a short-term tenant or licensee of an area of land owned by a state authority (Norman 1979, Budowski 1980). Such systems are fairly common throughout the humid zone of Africa. Descriptions of the Taungya system in Sierra Leone, Côte d'Ivoire, Togo, Nigeria, and

Kenya can be found in Koroma (1982), Tchoume (1982), Nadjombe (1982), Roche (1974) and Owino (1982), respectively.

The main advantage of this system is that it provides the shifting cultivators with fertile plots of land to cultivate while at the same time ensuring the regeneration of forests. Farming in this environment is a low-input system in which human labor is the predominant input. Most of the prospects for mechanization are in field-to-home and home-to-market transport. In the Taungya system the distance to the field affects productivity. Farmers who have to walk long distances to reach their fields may not be able to provide as much labor as farmers who rely on mechanical means of reaching them. Farmers in Trinidad, for example, used public transport to get to their Taungya fields, which were sometimes as far as sixteen kilometers from their homes. In Nigeria, farmers traveled as far as six kilometers by bicycle or motor vehicle to their farms (Enabor and others 1982).

As population density rises, agrosilviculture systems are economically feasible only in areas where other, more profitable, agricultural systems are unlikely to be established. The valley bottoms and alluvial plains, for instance, are unlikely candidates for agroforestry, since they are better suited to permanent cultivation of rice and vegetables. More marginal lands on the upper slopes and the steeper midslopes are the most likely areas for agrosilviculture. These areas are already under forest or bush cover, and permanent cultivation of field crops is not possible because of high levels of erosion and low soil fertility. The deeper soils of the mid and lower slopes can be cultivated with a combination of perennial and annual food crops and are less likely candidates for agrosilviculture.

In the agroforestry system, annual crops are grown, usually in mixtures, under perennial forest or plantation trees. The home gardens of South and Southeast Asia and the compound farms of West Africa are the best-known examples of agroforestry. Permanent compound farming represents an intensive management system that approximates the ecosystem of the forest. Lal (1983) points out that such a system can sustain and prolong the productivity of land and minimize degradation of the soil in the humid tropics even at high population density. In densely populated central Java, for instance, home gardens account for 22 percent of the total area under cultivation (Ruthenberg 1980).

Compound farms and home gardens are also found in several parts of Sub-Saharan Africa. Lagemann (1977) provides a description of the home gardens of Eastern Nigeria. Mixtures of annual crops, consisting of yams, maize, cassava, groundnuts, and vegetables, are planted between trees such as oil palm, breadfruit, kola nut, and mango. All trees are pruned or heavily trimmed to allow sunlight in; the cut branches are used for building material and firewood, while the small twigs and leaves are used as mulch and as feed for goats. Annual crops are directly seeded without any land preparation with hoe or plow. Lagemann found that the quantity of mulch and manure used and the density of the crop mixture increased with increasing scarcity of land.

The primary advantage of home gardens is that they can sustain high intensities of crop cultivation with minimum levels of soil degradation in the humid tropics. Perhaps, in recognition of this, the International Institute of Tropical Agriculture (IITA) in Nigeria has been promoting a system of farming known as alley cropping. This system, regarded as an improvement over traditional bush-fallow techniques, consists of planting selected species of trees and arranging them in such a way as to allow for narrow alleys in which food crops are grown. Experimental results indicate that sustainable yields are obtainable in what appears to be a stable system (Getahun, Wilson, and Kang 1982; Kang and Juo 1981).

Mechanization choices for home gardens and alley cropping are limited. Transport, both on and off the farm, and food processing continue to be important subjects for mechanization. Tillage requirements are minimal in these systems, so there should be no need for animal- or tractor-drawn plows. IITA uses mechanized techniques for pruning the trees, but the advantage of these techniques over hand pruning is limited in traditional, subsistence home gardens.

Systems without Trees

Where the markets for tree crops and forest products are limited and where rising population density leads to a reduction in fallow periods, the problem of finding a workable farming system for the low-CEC soils of the humid tropics appears. The problem is essentially that arable cropping patterns similar to those used in the temperate environment have adverse effects on the soil and therefore cause a rapid decline in yields (Lal 1983, Poulsen 1978, Kang and Juo 1981).

Experimental results at IITA indicate that intensive cultivation of the low-CEC soils of the humid tropics is sustainable only with a zero or minimum tillage system with a live mulch cover. This system involves the growing of a food crop, such as maize, directly in a living cover of a low-growing plant with minimum disturbance of the soil. It incorporates the soil-conservation features of organic mulch and no tillage but has the additional advantage of weed control. The live legume mulch also has the advantage of contributing nitrogen to the system (Kang and Juo 1981, Lal 1983). Experiments using crop residues as mulch have also been conducted at IITA; this system requires high levels of herbicides for weed control, however, and is therefore not practical for subsistence production (Lal 1983).

IITA has developed hand-pushed punch planters for no-till farming, which have not been adopted anywhere, presumably because the economic benefits of using the planter instead of a hand hoe are minimal. More recently, punch planters have been mechanized by attaching them to a power tiller or a small tractor. It is unlikely that the mechanization of the no-till seeding operation will be profitable. Once again, the operations that can profitably be mechanized are field-to-market transport and mechanical milling of food grains. Among the humid tropical locations that we have visited the mechanization of these operations is well under way.

Until this point we have concentrated on choices of farming systems for the upper and mid slopes of the humid tropics. These are the areas in which on most soils intensive cultivation of field crops is not possible. This is not true, however, in the valley bottoms, where intensive production of rice is technically feasible and is often economically attractive where population density is high or where markets exist. Mechanization choices in these instances have already been discussed. The potential for mechanized land preparation in these areas is great.

Chapter 12

Public Tractor-Hire and Equipment-Hire Services

Public tractor and equipment hire services have often been seen as an effective means of mechanizing agricultural operations without imposing high capital acquisition costs on the farmer. This expectation has generally not been fulfilled, and such attempts have usually failed.

In areas in which private rental markets are not feasible because of the time-bound, synchronic nature of the operation, a public hire service is even less of a possibility, because in addition to problems of timeliness, there is the added cost of a bureaucratic organization. In areas in which private rental markets are feasible, however, there is no need for a public tractor-hire service.

Here we shall summarize a review of a report entitled "Public Tractor and Equipment Hire Schemes in Developing Countries," by Peter Seager and Robert Fieldson (1984). They undertook this review as part of our research project on agricultural mechanization in Sub-Saharan Africa. Information was collected on twenty-one tractor-hire schemes throughout the world, fourteen of which were in Africa.

Background and Historical Perspective

Tractor and other machinery-hire services had their origin during World War II, when many European governments organized these efforts to stimulate agricultural production in a time of labor shortage. During the late 1940s and early 1950s, much effort went into increasing agricultural production in the then European colonies. This led to the setting up of tractor and implement hire services in the colonies, often under the direct supervision of a government organization. Some government organizations that provided machinery-hire services were the

Colonial Agricultural Machinery Advisory Committee of the United Kingdom, la Section de Mechanisation Agricole au Ministère d'Outre-Mer, France, and the Netherlands Advisory Board for the Mechanization of Agriculture in Overseas Territories. Subsequently, several of the newly independent countries continued the tractor-hire services, often expanding them to include the entire country or all crops.

The objectives or rationale for setting up a tractor-hire service are varied. The main objectives are to increase the production of export- and import-substitution crops; to facilitate speedy resettlement of population in new, sparsely populated areas, to demonstrate the use and benefit of a new technology, and to encourage the modernization of traditional agricultural systems. Of the twenty instances in which objectives for setting up the service were given, eighteen were equally divided between the first three objectives and two were pursuing the fourth. Let us briefly evaluate the prospects for success of each objective.

Export- and Import-Substitution Crop-Production Schemes

Many of the tractor-hire services are linked directly or indirectly to the introduction of a new crop or an increase in emphasis on an existing crop. These are either crops for export such as cotton or crops such as cereals, for import substitution. Cotton and rice were equally important—ten instances each—as the main crops among the twenty-one documented cases. The implicit rationale behind such schemes is that use of the tractor will lead to an expansion in the area planted in the crop in question and therefore increase total production of the crop. Did acreage planted in that crop increase because of the tractor-hire service? The answer is no in all six instances in which expansion in crop area was stated as an objective.

Land-Settlement Schemes

Land-settlement schemes were started in response to land pressure, often moving farmers from upland areas of intensive farming to savanna areas where it was hoped intensive mechanized systems with tractor-hire services could be used. Among countries in which there has been resettlement on account of land pressure are Kenya, Peru, the Central African Republic, and Belize. The success or failure of these schemes will not be apparent for a long time, since the move usually means a drastic and sudden change in farming systems, and the learning and settlement process takes a long time. While a case can be made for some degree of initial subsidy, it is not clear that the tractor-hire service is the most efficient way to use it. With heavy land-clearing and -shaping equipment, however, as in Peru and Belize, public hire of machinery does seem to function efficiently.

When new areas are being developed, roads are often inadequate and maintenance infrastructure limited, so overhead costs are high. Wear and tear on machinery will also be high, causing operating costs to be high as well. Further, heavy machinery is so expensive, private contractors are unlikely to become heavily involved. In these instances, there is a strong a priori case for public involvement, since the machinery can be operated efficiently and the "development" subsidy can be justified.

Demonstration

Demonstration schemes imply a gradual educative process rather than a summary dismissal of farming methods already in use. Tractor-hire services have been used in demonstration schemes in Swaziland, Malawi, Gambia, Nigeria, Egypt, and Malaysia. The tractor-hire service is usually attached to a station of the ministry of agriculture, demonstrations being carried out on station land, and these are linked to the overall extension effort. There has been some evidence of success in the demand for the tractor-hire service and subsequent initiation of private hire schemes, as in Malaysia and Egypt. Problems arise when subsidized services are continued beyond the introductory period, encouraging misallocation of resources by the farmers and slowing the entry of private contractors into the hire market. Government tractor-hire service remains the principal source of tractor service, and the farmers are thus restricted to a service that is usually unreliable. Some farmers do begin buying tractors for their own use and do provide a limited hire service, but they cannot compete with the government-subsidized rate.

Modernization Schemes

In some countries—Uganda, Malawi, and Kenya, for example—modernization of smallholder agriculture was the main objective. These schemes have been costly failures. The idea behind them was to induce a rapid change in the traditional farming system with extensive tractorization. Often tractors are introduced into areas of bush-fallow agriculture, where the costs of clearing the land and sustaining a tractor operation are extremely high. These schemes have failed because they were inappropriate and the demand was not great enough.

Operations for Which a Tractor-Hire Service Is Provided

Table 12-1 lists the twenty-one tractor-hire schemes reviewed, by operations performed. The operation for which the tractor-hire service was most commonly provided—in nineteen of the twenty-one examples—was land preparation. The moldboard plow is, in fact, the most common implement possessed by a tractor-hire service. Among the other cropping operations, transport was the next most common—although substantially less common than plowing; it was provided in five of the twenty-one examples. Mechanized planting and mechanized harvesting were attempted in a few instances—in three of twenty-one and two of twenty-one, respectively. Successful threshing operations were reported in Egypt, Sri Lanka, and Malaysia. In Egypt, for instance, threshing accounts for 30 percent of the total work time of tractors. Tractor-mounted sprayers are hired out in the Central African Republic, Togo, and Côte d'Ivoire for application of pesticides on cotton. Finally, in the densely populated areas of Malaysia and Bangladesh tractor-powered irrigation pumps are used to obtain a second crop of paddy; in Egypt this equipment is used for irrigating the primary crops of cotton, wheat, and maize.

Table 12-2 shows examples of the distribution of time, by operation, per tractor. This table shows that even in instances in which operations other than plowing are provided for by the hire service, plowing is the one that is most important. The time spent in plowing is around 90 percent of the total work time of the tractor and equipment in all examples except that of Egypt, which has successfully diversified its hire service to include threshing, transport, and irrigation.

Overriding Problems of the Tractor-Hire Service

Any public tractor-hire service suffers from one or more of the following problems, which severely affect its success: inappropriate farming intensity; time-bound, synchronic operations; the need to provide incentives for operators; and lack of financial autonomy. Each problem will be discussed in detail.

Evolution of Farming Systems and Tractor-Hire Services

A necessary condition for the success of a tractor-hire scheme is that the farming system in the area be at an intensity sufficient to warrant

Table 12-1. *Use of the Tractor-Hire Service for Various Farming Operations in Twenty-one Countries Surveyed*

Variable	*Land clearing*	*Land preparation*	*Planting*	*Harvesting*	*Threshing*	*Transport*	*Spraying*	*Irrigation*
Number of instances	7	19	3	2	3	5	3	2
Percent of instances	33	90.5	14	9.5	14	23.8	14	9.5
Countries	Benin Sierra Leone, Ghana, Kenya, Pakistan, Belize	All except Central African Rep. and Pakistan	Ghana, Kenya, Swaziland	Ghana, Belize	Egypt, Sri Lanka, Malaysia	The Gambia, Côte d'Ivoire, Swaziland, Bangladesh, Nicaragua	Central African Rep., Togo, Côte d'Ivoire	Malaysia, Bangladesh

Source: Seager and Fieldson (1984).

Table 12-2. *Percentage of Time Spent by the Tractor-Hire Service on Each Farming Operation, Selected Instances Out of Twenty-one Surveyed*

Country	*Land clear-ing*	*Land prep-aration*	*Planting*	*Inter-culture*	*Spray-ing*	*Harvest-ing*	*Thresh-ing*	*Trans-port*	*Irri-gation*
Ghana	4	91	2			3			
Kenya		93	7						
Egypt									
Government		52					29	17	2
Private		40					39	15	4
Swaziland		90	10						
Uganda		90			2.8				

Source: Seager and Fieldson (1984).

plowing with tractors. Attempts to introduce tractor-hire services into forest-fallow and bush-fallow systems will invariably fail because of the high density of stumps, which leads to high operating and repair costs, and the extensive nature of the farming system, which implies low levels of demand for land preparation by means of the tractor-drawn plow. It should be noted that contract-hire operations, whether public or private, will fail under low farming intensity.

Synchronic and Time-Bound Operations

Much of the problem regarding tractor-hire services is that the operation to which the tractor is best suited, plowing, is time-bound, usually needing to be done within a short period, and synchronic—needing to be done within a particular region at the same time on all farms. Rental markets are not generally successful under such conditions because of severe conflicts among potential users as to the sequence in which they are to be served. There is no acceptable managerial solution to this problem. Further, public hire schemes suffer from a number of problems that are directly attributable to poor management. First, the percentage of nonproductive working time is high in connection with multifarm use because of travel time. Second, with public tractor-hire services the beginning of the peak season often finds many of the tractors still out of working order. Finally, at the peak season, when the machines should be working around the clock, public tractor-hire services are often held back by incentive problems with operators who will work only the regulation eight-hour day. Some farmers therefore choose private contract-hire operators, when such are available, in spite of the higher unit costs of their services.

Where the tractor service is used primarily for plowing, rates of utilization are often extremely low because of the time-bound and synchronic nature of the operation. Tractor-utilization rates can be improved if the tractors are allowed to move across agroclimatic zones. Kenya initially took this into account in its design of the tractor-hire service and achieved high rates of utilization of its tractors.

The government tractor-hire service in Kenya originated in 1966–67, when fifty tractors were introduced to provide a cultivation service, primarily for areas under the Masai Wheat Scheme. Operating in mobile units with no permanent bases, the tractors could be moved from area to area to follow the work. With Kenya's diverse ecology and agriculture, virtual year-round utilization of machinery was possible. In 1967–68, for example, more than 1,000 hours per tractor were recorded, with an average of 772 productive hours per machine; the proportion of productive hours to total running hours was thus about 70 percent.

In 1980, the mobile nature of the operation was changed and the tractors were assigned to specific stations from which they were to operate. The total number of tractors available for contract-hire rose from 50 to 150, yet the number of productive hours worked per tractor decreased to 167 a year. Basing tractors at specific locations thus led to a steep decline in the rate of utilization of the equipment.

The harvesting operation is similarly time-bound and generally not provided by a tractor-hire service, although Belize attempted to supply rice-harvesting services. In Belize 90 percent of all mechanized rice is of the same variety, and it all germinates at the same time, so there is a maximum of four to six weeks for the harvest. The logistical problems have caused rates of utilization to be only around eighty hours per machine a year.

Threshing, on the other hand, is not time-bound and can be provided by contract-hire operators. In Sri Lanka, Egypt, and Malaysia tractor-powered threshers were provided as part of the tractor-hire service. In Egypt, for instance, 60 percent of all wheat and 50 percent of all rice is threshed by tractor-powered drum threshers, and more than half the threshing is done by the government tractor-hire service.

Transport does not suffer from any of the timeliness problems and can easily be provided on a contract-hire basis. In only five of the twenty-one instances reviewed, however, was this service provided—probably because tractor-hire for transport competes with several other transport systems, such as trucks, bush taxis, and buses, which may be providing a cheaper or speedier service.

Clearing and leveling of the land is the one activity for which equipment-hire services have been successfully established. Public bulldozer services have been provided in Benin, Sierra Leone, Ghana, Kenya, Pakistan, Peru, and Belize. In Pakistan, for instance, the provincial governments provide fleets of bulldozers and associated equipment for land-development work. The machines are hired out for new land development—clearing bush and trees—and for leveling already cultivated land. The rate of utilization of this equipment is quite high, each machine averaging about 1,500 hours a year.

Incentive Problems

In addition to the time-bound nature of the plowing operation, another reason for the low utilization rate of equipment in a public-hire service is that the operators have no incentive to work beyond the minimum prescribed hours. The staff of most hire services are employees of the ministries of agriculture, are paid regular government salaries, and have considerable employment security. Since salaries are often not tied to number of hours worked there is no incentive to work longer hours, even in the peak season. This problem is encountered in Sierra Leone, Malawi, and Kenya, among other countries.

A few countries have tried to provide incentives to operators by offering overtime payments or some other unit payment for work done, and the result has generally been higher work rates. In Swaziland, for example, the operators were kept on all year until 1979, but they were given generous overtime allowances at peak periods. Once these allowances were stopped the performance of the operators dropped sharply. In Pakistan, the staff of the hire service are paid regular salaries according to government pay scales, but it is usual for customers to make an additional hourly payment to drivers. In addition, awards are made to the drivers and mechanics of the machines that achieve the best performance. Without payment or bonus per unit of work the operators have no incentive to work the maximum number of hours during the peak season.

Financial Success

Do public tractor-hire services operate at a profit or at least break even? The answer is a definite no in all cases. Table 12-3 indicates the level of subsidy provided in selected instances. The subsidy ranged from 94 percent in Sierra Leone to 24 percent in Swaziland. Why do public tractor-hire services operate at a loss? The price for the service is decided by a central authority without consideration of local demand conditions; the overhead of the hire service is very high—tractor-hire stations in Kenya, for example, had as many as 100 employees; rates of utilization are extremely low; and it is politically unappealing to withdraw the subsidy once it has been established. It should be emphasized here that private tractor-hire services have generally operated at a profit. The bottom line for public tractor hire services is: Where private rental markets are not feasible, tractor-hire services are even less feasible because in addition to problems of timeliness there is the added cost of a bureaucratic organization. Where private rental markets are feasible no public tractor-hire service is needed.

Can Private Tractor-Hire Services Do Better?

The primary defects of the public tractor-hire services are that they have an extremely low rate of utilization and that they operate at a loss.

Table 12-3. *A Comparison of Government and Private Tractor-Hire Services in Selected Countries, Selected Years, 1967–82*

		Government				Private	
Country	Year	Number of hours of use a year	Cost per hectare[a]	Revenue per hectare[a]	Subsidy (percent)	Number of hours of use a year	Net revenue per hectare[a]
Nigeria	1972	300	n.a.	13.00	n.a.	800	n.a.
Nigeria[b]	1981–82	400	7.20	5.00	28	n.a.	n.a.
Kenya	1981	200			60	1,000	n.a.
Egypt	1978	410	6.40	2.00	69	1,534	0.66
Sierra Leone	1982	324	n.a.	n.a.	94	n.a.	n.a.
Swaziland	1978–79	738[c]	8.43	6.50	24	n.a.	n.a.
Sudan	1979–80	171	n.a.	n.a.	67	n.a.	n.a.
Uganda	1968	250	n.a.	n.a.	45	800	n.a.
Malawi	1967	130	11.67	6.00	49	n.a.	n.a.

n.a. Not available.

a. Cost figures are in local currency.

b. Figures from Anazodo (1983).

c. 300 hours per hectare is spent on departmental work of the Rural Development Area, including the grading of roads, transport of materials, and demonstration.

Source: Seager and Fieldson (1984).

In table 12-3 the utilization rates of public and private tractors provided on hire are compared. Where these data were provided private tractors worked more hours a year than public tractors. This is not surprising, since operators of private tractors have an incentive to work as many hours as they can during the peak season. Also, private contractors are not restricted to a particular area and can increase their rates of utilization by moving among several agroclimatic zones. This practice is common among the tractor-hire operators in Kenya. We do not have cost and revenue figures for private operations, except those in Egypt, but we could expect a priori that private operations would be more profitable because of lower overhead and higher rates of utilization.

The case of Egypt illustrates clearly the difference between private and public hire services. Tractor-hire services in Egypt are provided by private owners and by a cooperative service. Table 12-4 presents a comparison of cooperative and private tractors. The main difference is in the total hours of annual use—410 by cooperatives, 1,534 by private. Of all the cooperatively owned tractors surveyed, 22 percent were not in operation, while only 3 percent of those owned privately were idle (Egypt 1978). The private tractors operate at a profit, while the cooperative tractors must be subsidized. Low utilization of farm machinery

Table 12-4. *A Comparison of the Cost of Tractor Work Done by Private Enterprise with That Done by Government Enterprise and with the Actual Rate of Hire, Egypt*

Variable	*Agricultural cooperatives*	*Government enterprise*	*Private enterprise*
Number of tractors	30	320	20
Average number of hours worked a year	410	690	1,534
		Egyptian pounds	
Depreciation	2.43	1.04	0.65
Fuel and oil	0.40	0.71	0.62
Maintenance and repair	2.16	1.62	0.65
Operators	0.20	0.20	0.42
Overhead	1.21	1.04	n.a.
Total cost	6.40	4.61	2.34
Actual rate of hire	2.00	2.00	3.00
Amount of subsidy	4.40	2.61	0.00
Profit			0.66

n.a. Not available.
Source: Seager and Fieldson (1984, 99).

and equipment and uneconomic charges, especially for primary tillage, have led to substantial losses by many cooperatives.

The performance of a private hire service cannot be accurately measured while a subsidized public service is available, since the latter effects the functioning of the former, especially in the setting of a price. Take the extreme instance of Sierra Leone: with a subsidy of 96 percent it would be impossible for any private contractor to compete with the public service. Even in Swaziland, where the subsidy is 24 percent, private contractors are reluctant to enter the market, although the public tractor-hire service does not meet the demand for tractor services. In a recent tractor project in Uganda, machines were sold on credit to farmers who achieved utilization rates of 800 to 1,500 hours per machine a year. These examples indicate that in the absence of distortions, there may be great potential for setting up a private contract-hire service.

Conditions Necessary for Successful Contract-Hire Operations

The successful contract-hire operation must be private. Decisions on the operations to be performed, prices to be charged, and so on, are better made by individual contractors than by a centralized government authority. Individual contractors, moreover, would not lack incentives, since to them higher utilization rates mean higher revenues.

Land-clearing services for new land settlements, road building, and the like, could conceivably be provided by a government scheme. Land clearing requires specialized equipment—the bulldozer—which cannot be used for any other operation. In a given geographic area land-clearing operations are of a one-time nature—that is, the demand for the service is high during the period of settlement, then diminishes. For that reason individual contractors may be unwilling to invest in this equipment. The government organization, on the other hand, can move the equipment from one region to another as the need arises, thereby sustaining a high rate of utilization.

Contract-hire operations for land preparation should be mobile, capable of traveling across agroclimatic zones in order to achieve high rates of utilization. By being mobile the service can overcome the time-bound and synchronic nature of the land-preparation operation.

References

Abeles, Marc. 1981. "In Search of the Monarch: Introduction of the State among the Gamo of Ethiopia." In *Modes of Production in Africa: The Precolonial Era,* ed. Donald E. Crummey and Charles C. Stewart. London: Sage Publications.

Adam, F. H., and others. 1981. "Problems of Mechanized Agriculture in Developing Countries: The Case of the Central Rainlands of the Sudan." *Development and Peace* 1 (2).

Agabawi, K. A. 1968. "The Rain Grown Sorghum Development, Sudan Republic," in *Change in Agriculture,* ed. A. H. Bunting. London: Gerald Duckworth & Co.

Ahn, Peter M. 1970. *West African Soils.* London: Oxford University Press.

Allan, G. R. 1983. "Development of the Mumias Sugar Company, Kenya," *Oxford Agrarian Studies* 12.

Allan, William. 1965. *The African Husbandman.* Edinburg: Oliver and Boyd.

Almy, Susan W. 1979. "The Response of Agricultural Systems to Natural Increase: The History of a High-Potential Region in Kenya." In *Changing Agricultural Systems in Africa,* ed. Emilio F. Moran. Studies in Third World Societies, no. 8. Williamsburg, Va.: College of William and Mary. 1979.

Anazodo, U. G. N. 1983. "Investigative Surveys on Farm Tractors and Equipment Hiring Services in Nigeria." Nsukka: University of Nigeria Department of Energy. Processed.

Anthony, K. R. M., Bruce F. Johnston, W. O. Jones, and V. C. Uchendu. 1979. *Agricultural Change in Tropical Africa.* Ithaca: Cornell University Press.

Anthony, K. R. M., and V. C. Uchendu. 1970. "Agricultural Change in Mazabuka District, Zambia." *Food Research Institute Studies* 9:215–67.

———. 1974. *Agricultural Change in Geita District, Tanzania.* Nairobi: East African Literature Bureau.

Baier, Stephen. 1980. *An Economic History of Central Niger.* Oxford: Clarendon Press.

Barker, Randolph, William H. Meyers, Cristina M. Crisostomo, and Bart Duff. 1973. "Employment and Technological Change in Philippine Agriculture." In *Mechanisation and Employment in Agriculture: Case Studies from Four Continents.* Geneva: International Labour Office.

Barrett, Vincent, Gregory Lassiter, David Wilcock, Doyle Baker, and Eric Crawford. 1982. "Animal Traction in Eastern Upper Volta: A Technical, Economic and Institutional Analysis." International Development Paper no. 4. East Lansing: Michigan State University.

Bartsch, William H. 1977. *Employment and Technology Choice in Asian Agriculture.* New York: Praeger Publishers.

Bernard, Frank Edward. 1972. *East of Mount Kenya: Meru Agriculture in Transition.* Munich: Weltforum Verlag.

Bernsten, R. H. 1981. "Effects of Mini-Tractor Mechanization on Employment and Labor Use Intensity, Sidrap and Pinrang, South Sulawesi, Indonesia." Paper presented at a Workshop on the Consequences of Small Rice Farm Mechanization in Asia, 14–18 September. Los Baños, Laguna, Philippines: International Rice Research Institute (IRRI).

Bigot, Yves. 1981. "The Constraints to the Evolution of Animal Traction in the Savanna Zone of Ivory Coast" (in French). *Cashiers du CIRES* 30 (September): 9–29.

Binswanger, Hans P. 1978. "The Economics of Tractors in South Asia." ADC/ICRISAT Monograph. Hyderabad: International Crops Research Institute for the Semi-Arid Tropics (ICRISAT).

———. 1984. *Agricultural Mechanization: A Comparative Historical Perspective.* Staff Working Paper no. 673. Washington, D.C.: World Bank.

Binswanger, Hans P., and John McIntire. 1986. "Behavioral and Material Determinants of Production Relations in Land Abundant Tropical Agriculture." *Journal of Economic Development and Cultural Change* Forthcoming.

Binswanger, Hans P., and Prabhu L. Pingali. 1984. "The Evolution of Farming Systems and Agricultural Technology in Sub-Saharan Africa." Working Paper no. ARU 23. Washington, D.C.: World Bank, Agriculture and Rural Development Department, Research Unit.

Binswanger, Hans P., and Vernon Ruttan. 1978. *Induced Innovation: Technology, Institutions and Development.* Baltimore: Johns Hopkins University Press.

Bloom, Roger A. 1979. "A Review of the Technical Evidence on the Use of Animal Traction in Sahelian Farming Systems." East Lansing: Michigan State University, Department of Agricultural Economics. Processed.

Bonnefond, P. H. 1967. *Problems of Profitability of an Agricultural Development Action of the SATEC Operation in Mossi* (in French). Bondy, France: Ministère de Plan: Office de la Recherche Scientifique et Technique Outre-Mer (ORSTROM).

———. 1969a. "Introduction of Motorization in Ivory Coast Traditional Agriculture" (in French). Bondy: ORSTROM. Processed.

———. 1969b. "The Development of Agricultural Motorization in the Ivory Coast" (in French). Bondy: ORSTROM. Processed.

Boserup, Ester. 1965. *Conditions of Agricultural Growth.* Chicago: Aldine Publishing Co.

———. 1981. *Population Growth and Technological Change: A Study of Long-Term Trends.* Chicago: University of Chicago Press.

Botswana, Government of. 1980. "Dryland Farming Research Scheme (DLFRS) Botswana. Phase III. First Annual Report." Gaborone: Ministry of Agriculture, Division of Agriculture Research.

Bourn, David. 1983. "Tsetse Control, Agricultural Expansion and Environmental Change in Nigeria." Ph.D. dissertation, Christ Church College, Oxford University.

Budowski, G. 1980. "Agro-Forestry in Managing Tropical Forests." Paper presented at the International Symposium on Tropical Forests, Yale University, 15–16 April.

Burch, D. 1974. "Agricultural Tractor Pools in Ceylon 1952-1956." Unpublished Seminar Paper, University of Sussex.

Carr, M. N. 1975. "Patterns of Tractorisation in the Major Rice Growing Areas of Sri Lanka." Ph.D. dissertation, University of Sussex.

Carr, Stephen. 1982. "The Impact of Government Intervention on Smallholder Development in North and East Uganda." Occasional Paper no. 5, Agrarian Development Unit, School of Rural Economics, Wye College, University of London.

Carter, L. M., J. R. Stockton, J. R. Tavernetti, and R. F. Colwick. 1965. "Precision Tillage for Cotton Production." *Transactions of the American Society of Agricultural Engineers* 8:177–79.

Casse, M., Dumas, and Garin. 1965. "Results of Cultivation Trials in Francophone West Africa" (in French). Vols. 1, 2 and 3. Paris: Ministère de la Coopération.

———. 1969. "SEAE Republic of Togo: A Study of the Evolution of Production Factors Implemented during the Past 10 Years and Their Effect" (in French), Anthony, France.

CEEMAT (Centre d'Études et d'Experimentation du Mécanisme Agricole Tropical). 1972. "SEAE Republic of the Ivory Coast: A Study of the Evolution of the Factors of Production Implemented during the Past 10 Years and Their Effects" (in French). Anthony, France.

———. 1979. Mission au Nigeria. "Evaluation of the System of Agricultural Service Centers" (in French). Anthony, France.

Chancellor, W. J. 1973. "Mechanisation of Small Farms in West Malaysia." *Malaysian Agricultural Journal* 48 (2).

Charreau, Claude. 1974. "Soils of Tropical Dry and Dry-Wet Climatic Areas of West Africa and Their Use and Management." A series of lectures presented at Cornell University, Ithaca, N.Y.

Charreau, Claude, and R. Nicou. 1971. "The Improvement of the Cultural Profile in the Sandy and Clayey Soils of the Tropical Zone of Southwest Africa and Their Agronomical Occurrence" (in French). *L'Agronomie tropicale,* no. 9.

Clayton, Eric S. 1974. "Mechanisation and Employment in East African Agriculture." *International Labor Review.* 110 (1).

Colson, Elisabeth. 1962. "Trade and Wealth among the Tonga." *In Markets in Africa,* ed. Paul Bohannan and George Dalton. Evanston, Ill.: Northwestern University Press.

Côte d'Ivoire, Ministère de L'Agriculture. 1966–70. Project for the Motorization of the Agriculture of the Ivory Coast.

Curfs, H. P. F. 1974. "Soil Preparation and Weed Control for Upland and Irrigated Rice Growing." In Report on the Expert Meeting on the Mechanization of Rice Production, International Institute of Tropical Agriculture (IITA), Ibadan, Nigeria, 10–14 June.

Curtin, Philip, Steven Feierman, Leonard Thompson, and Jan Vansina. 1978. *Africa History.* Boston: Little, Brown & Co.

Day, R. H., and I. J. Singh. 1977. *Economic Development as an Adaptive Process: The Green Revolution in the Indian Punjab.* Cambridge: Cambridge University Press.

Delgado, Christopher L. 1979a. "Livestock versus Food Grain Production in Southeastern Upper Volta: A Resource Allocation Analysis." Ph.D. dissertation, Cornell University, Ithaca, N.Y.

———. 1979b. *The Southern Fulani Farming System in Upper Volta: A Model for the Integration of Crop and Livestock Production in the West African Savannah.* African Rural Economy Paper no. 20. East Lansing: Michigan State University.

Delgado, Christopher L., and John McIntire. 1982. "Constraints on Oxen Cultivation in the Sahel." *American Journal of Agricultural Economics* 64 (2) (May):188–96.

Desai, D. K., and C. Gopinath. 1975. *Impact of Farm Tractorization on Productivity and Employment (Gujarat State).* Ahmedabad: Center for Management in Agriculture, Indian Institute of Management.

Desmond, P. 1983. Final Report on Termination of the Mechanical Workshop Project. Government of Sierra Leone, Ministry of Agriculture and Forest.

De Vletter, F. 1979. "Socio-Economic Profile of Swazi Rural Homesteads." Mbabane: University of Swaziland.

Dudal, R. 1980. "Soil-Related Constraints to Agricultural Development in the Tropics," In *Soil Related Constraints to Food Production in the Tropics.* Los Baños, Laguna, Philippines; International Rice Research Institute.

Egypt, Government of. 1978. *Farm Machinery Survey.* Cairo: Ministry of Agriculture.

Eicher, C. K., and Doyle Baker. 1982. *Research on Agricultural Development in Sub-Saharan Africa: A Critical Survey.* International Development Paper no.

1. East Lansing: Michigan State University.

El Hosery, Ali M. 1982. "Evaluation of Multi-Farm Use of Agricultural Machinery in Egypt." DLG. Mechanization Symposium.

El-Sharhawy, M. A., and K. Sgaier. 1975. "Effect of Tillage on Root Penetration and Grain Yield of Mexican Wheat Grown in the Libyan Desert." *Libyan Journal of Agriculture* 4:25–28.

Enabor, E. E., J. A. Okojie, and I. Verinumbe. 1982. "Taungya Systems: Socio-Economic Prospects and Limitations." In *Agro-Forestry in the African Humid Tropics,* ed. L. H. MacDonald. Tokyo: United Nations University.

Engelhardt, Thomas. 1984. "Economics of Traditional Smallholder Irrigation Systems in Semi-Arid Tropics of South India." Ph.D. dissertation, Universität Hohenheim, Stuttgart.

Evenson, Robert E. 1982. "Government Policy and Technological Progress in U.S. Agriculture." In *Government Support of Technological Progress: A Cross Industry Analysis,* ed. Nelson Richard. New York: Pergamon Press.

FAO (Food and Agriculture Organization of the United Nations). 1975. "Farm Mechanisation in the Republic of Sri Lanka." Rome: Agricultural Engineering Division. Processed.

———. 1983a. *Animal Health Yearbook, 1982.* Animal Production and Health Series, no. 19, Rome.

———. 1983b. *African Animal Trypanosomiasis: Selected Articles from the World Animal Review.* Animal Production and Health Series, no. 37. Rome.

———. 1984. *Institutional Aspects of Shifting Cultivation in Africa.* Rome: Human Resources, Institutions and Agrarian Reform Division.

FAO/CEEMAT. 1972. *Manual on the Employment of Draught Animals in Agriculture.* Rome.

Farrington, John. 1985. "Farm Mechanization in South Asia: The Formulation and Impact of Recent Policies." Paper presented at the IRRI Workshop on Small Farm Equipment for Developing Countries: Past Experience and Future Priorities. International Rice Research Institute, Manila, Philippines, 2–7 September.

Farrington, John, and F. Abeyratne. 1982. *Farm Power in Sri Lanka.* Development Study no. 22. Department of Agricultural Economics and Management, University of Reading.

Faucher, D. 1951. *L'Homme et la machine* [Man and Machine]. Paris: Édition de Minuit.

Ford, J. 1971. *The Role of Trypanosomiasis in African Ecology, A Study of the Tsetse Fly Problem.* Oxford: Clarendon Press.

Gambia, The, Ministry of Agriculture. 1972. "Tractor Ploughing Scheme." Mechanical Engineering/Transport Division.

———. 1981a. "Mechanisation of Rice Growing: An Investigation of Current Fixed Charges by the Rice Mechanisation Scheme." Draft to Permanent Secretary.

———. 1981b. A Review of Mechanisation, Land Use, and Tractor Ploughing in The Gambia. Mechanical Engineering/Transport Division.

Geddes, A. W. M. 1972. "Mechanical Cultivation Policy." Government of Sierra Leone, Ministry of Agriculture and Forest, Agricultural Economics.

Gemmill, G. T. 1971. *Tractor-Hire Services: A Case Study of Malawi.* Bunda College of Agriculture, University of Malawi.

Getahun, A., G. F. Wilson, and B. T. Kang. 1982. "The Role of Trees in Farming Systems in the Humid Tropics." In *Agro-forestry in the Humid Tropics,* ed. L. H. MacDonald. Tokyo: United Nations University.

Ghana, Ministry of Agriculture. 1983. "Agricultural Mechanisation in Ghana."

Gleave, M. B., and H. P. White. 1969. "Population Density and Agricultural Systems in West Africa." In *Environment and Land Use in West Africa,* ed. M. F. Thomas and G. W. Whittington. London: Methuen & Co.

Greenland, D. J. 1979. "Structural Organization of Soils and Crop Production." In *Soil Physical Properties and Crop Production in the Tropics,* ed. R. Lal and D. J. Greenland. New York: John Wiley & Sons.

Greenland, D. J., and R. Lal. 1979. "Towards Optimizing Soil Physical Characteristics for Sustained Production from Soils in the Tropics." In *Soil Physical Properties and Crop Production in the Tropics.* ed. R. Lal and D. J. Greenland. New York: John Wiley & Sons.

Grimes, D. W., R. J. Miller, and P. L. Wiley. 1975. "Cotton and Corn Root Development in Two Field Soils of Different Strength Characteristics." *Agronomy Journal,* 67:519–23.

Grove, A. T. 1961. "Population and Agriculture in Northern Nigeria." In *Essays on African Populations,* ed. K. M. Barbour and R. M. Prothero. London: Routledge & Kegan Paul.

Hailey, Lord M. 1957. *An African Survey.* Revised edition. London: Oxford University Press.

Haudricourt, A. G., and Jean Brunhes Delamare. 1955. *Man and the Plow around the World* (in French). Paris: Gallimard, NRF (*Geographic Humaine* 25).

Hayami, Yujiro, and Vernon Ruttan. 1985. *Agricultural Development: An International Perspective.* Revised and expanded. Baltimore: Johns Hopkins University Press.

Henderson, D. H., and W. Ghericke. 1983. "Economics of Tractor Operation in Swaziland." Proceedings of the American Society for Agricultural Engineering.

Herdt, R. W. 1983. "Mechanization of Rice Production in Developing Asian Countries: Perspective, Evidence and Issues." In *Consequences of Small Farm Mechanization.* IRRI/ADC (Agriculture Development Council) conference volume. Los Baños, Laguna, Philippines: International Rice Research Institute.

Herskovits, Melville J. 1926. "The Cattle Complex in East Africa." *American Anthropologist* 28.

Hopkins, A. G. 1973. *An Economic History of West Africa.* New York: Columbia University Press.

Humphrey, Norman. 1945. *Thoughts on the Foundations of Future Prosperity in the Kikuyu Lands: Based on a Review of Some Factors Affecting Agriculture in the South Nyeri District.* Nairobi: Colony and Protectorate of Kenya.

———. 1947. *The Liguru and the Land.* Nairobi: Colony and Protectorate of Kenya.

Hurt, R. Douglas. 1982. *American Farm Tools: From Hand Power to Steam Power.* Manhattan, Kan.: Sunflower University Press.

Hyden, Gordon. 1980. *Beyond Ujamaa in Tanzania.* Berkeley: University of California Press.

Igbeka, I. G. 1983. *Economic Evaluation of Tillage Operation in Some Mechanized Farms in Nigeria.* Ibadan, Nigeria: University of Ibadan, Department of Agricultural Engineering.

Iliffe, John. 1979. *A Modern History of Tanganyika.* Cambridge: Cambridge University Press.

India, Government of. 1975. "Statistical Abstract of India 1975." Central Statistical Organization.

———. 1976. "Agricultural Situation in India." New Delhi: Directorate of Economics and Statistics.

Ing, V. Cervinka. 1967. "Tropical Farm Mechanisation: Ghana's Experience." *World Crops* 19.

Jabbar, M. A., M. S. R. Bhuiyan, and A. K. M. Bari. 1983. "Causes and Consequences of Power Tiller Utilization in Two Areas of Bangladesh," In *Consequences of Small Farm Mechanization.* IRRI/ADC Conference Volume. Los Baños, Laguna, Philippines: International Rice Research Institute.

Jaeger, William K. 1986. *Agricultural Mechanization: The Economics of Animal Traction in Burkina Faso.* Boulder, Colo.: Westview.

Jaetzold, Ralph, and Helmut Schmidt. 1982. Farm Management Handbook of Kenya. Vol. 2A. West Kenya. Nairobi: Government of Kenya, Ministry of Agriculture.

Jahnke, Hans E. 1976. *Tsetse Flies and Livestock Development in East Africa.* Munich: Weltforum Verlag.

———. 1982. *Livestock Production Systems and Development in Tropical Africa.* Kiel: Kieler Wissenschafts Verlag Vauk.

Jedrej, M. C. "The Growth and Decline of the Mechanical Agriculture Scheme in West Africa." *African Affairs.*

Jerve, Alf Morten. 1982. *Livestock Trypanosomiasis and the Pastoral Crisis in Semi-Arid Africa, an Approach to Future Land Use, with Example from Somalia.* DERAP (Development Research and Action Programme) Working Papers. Bergen, Norway: Chr. Michelsen Institute.

Johnson, C. E. 1956. *African Farming Improvement in the Plateau Tonga Maize Areas of Northern Rhodesia.* Agricultural Bulletin no. 11. Lusaka: Government of Northern Rhodesia, Department of Agriculture.

Johnston, Bruce F. 1981. *Farm Equipment Innovations and Rural Industrialisation in Eastern Africa: An Overview.* World Employment Programme Research, Working Paper WEP 2022/WP. 80. Geneva: International Labour Office.

———. 1984. "Farm Equipment Innovations in Eastern Africa: Policy Considerations," In *Farm Equipment Innovations in Eastern and Central Southern Africa,* ed. Iftikar Ahmed and Bill H. Kinsey. Geneva: International Labour Office.

Jones, William O. 1959. *Manioc in Africa.* Stanford, Calif.: Stanford University Press.

Jones, William I., and Roberto Egli. 1984. "Farming Systems in Africa: The Great Lakes Highlands of Zaire, Rwanda and Burundi." Technical Paper no. 27. Washington, D.C.: World Bank, 1984.

Joy, J. L., ed. 1960. Symposium on Mechanical Cultivation in Uganda, Department of Agriculture, Kampala.

Juarez, F., and R. Pathnopas. 1983. "Comparative Analysis of Thresher Adoption and Use in Thailand and the Philippines." In *Consequences of Small Farm Mechanization.* IRRI/ADC Conference Volume. Los Baños, Laguna, Philippines: International Rice Research Institute.

Kahlon, A. S. 1975. *Impact of Mechanization on Punjab Agriculture with Special Reference to Tractorization.* Department of Economics and Rural Sociology, Punjab Agricultural University, Ludhiana, India.

Kang, B. T., and S. R. Juo. 1981. "Management of Low Activity Clay Soils in Tropical Africa for Food Crop Production." Paper presented at the 4th International Soil Classification Workshop, Kigali, Rwanda, 2–12 June.

Kenya, Ministry of Agriculture and Animal Husbandry. 1967a. "Tractor Hire Service."

———. 1967b. "THS Report and Evaluation. 1966/67."

———. 1968. "THS Report and Evaluation. 1967/68."

———. 1981. "Tractor Hire Service Annual Report."

Kinsey, B. H. 1984. "Farm Equipment Innovations, Agricultural Growth and Employment in Zambia." In *Farm Equipment Innovations in Eastern and Central Southern Africa,* ed. Iftikhar Ahmed and Bill H. Kinsey. Hampshire: Grover.

Kjoerby, Finn. 1983. *Problems and Contradictions in the Development of Ox-Cultivation in Tanzania.* Uppsala: Scandinavian Institute of African Studies.

Kline, C. K.; D. A. G. Green; Roy L. Donahue, and B. A. Stout. 1969. *Agricultural Mechanization in Equatorial Africa.* College of Agriculture and Natural Resources Research Report no. 6. East Lansing: Michigan State University.

Kolawole, M. I. 1972. "Economic Aspects of Tractor Contracting Operations in Western Nigeria." *Journal of Agricultural Engineering Research* 17: 289–94.

———. 1974. "Economic Aspects of Private Tractor Operations in the Savanna Zone of Western Nigeria." *Journal of Agricultural Engineering Research* 19(4): 401–10.

Koroma, A. P. 1982. "Taungya in Sierra Leone." In *Agro-Forestry in the African Humid Tropics,* ed. L. H. MacDonald. Tokyo: United Nations University.

Kowal, J. M., and A. H. Kassam. 1978. *Agricultural Ecology of Savanna: A Study of West Africa.* Oxford: Clarendon Press.

Labrousse, G. 1961. "General Conditions of Development of Agricultural Mechanization in Tropical Countries, d'Expression Française." *Bulletin de CMAOM* 28: 6–11; 29: 9–15.

———. 1971. *Historic-Economic Aspects of the Development of Mechanized Cultivation of Food Crops in Tropical Countries.* Bambey, Senegal: Agricultural Research Center.

Lagemann, Johannes. 1977. *Traditional African Farming Systems in Eastern Nigeria.* Munich: Weltforum Verlag.

Lal, R. 1983. *No-Till Farming: Soil and Water Conservation and Management in the Humid and Sub-Humid Tropics.* Monograph no. 2. Ibadan, Nigeria: International Institute of Tropical Agriculture.

Lassiter, Gregory C. 1981. Cropping Enterprises in Eastern Upper Volta. African Rural Economy Paper no. 35. East Lansing: Michigan State University.

Ledgard, R. 1982. "A Case Study of Multifarm Use of Farm Machinery in the Central Huallago Region of Peru." Silsoe, England: National Institute of Agricultural Engineering. Processed.

LeHouerou, H. N., and G. F. Popov. 1981. *An Eco-Climatic Classification of Inter-Tropical Africa.* FAO Plant Production and Protection Paper no. 31. Rome: FAO.

Lele, Uma. 1975. *The Design of Rural Development: Lessons from Africa.* Baltimore: Johns Hopkins University Press.

LeMoigne, Marc. 1974. "Some Aspects of Tillage in Rice Culture." In Report on the Expert Meeting on the Mechanization of Rice Production, International Institute of Tropical Agriculture (IITA), 10–14 June, Ibadan, Nigeria.

LeMoigne, M., and D. Zerbo. 1977. *Problems Posed by Mechanization in the Member Countries of the C.I.L.S.S.* (Comité Permanent Intre-Etats de Lutte contre le Secheresse dans le Sahel). Antony, France: Centre d'Études et d'Experimentation du Mécanisme Agricole Tropical (CEEMAT).

Lhoste, P. 1983. "Development of Animal Traction and the Evolution of Pastoral Systems in Sine Saloum, Senegal" (in French). *Revue d'Élevage et de médécine vétérinaire des pays tropicaux* 36 (3): 291–300.

Ludwig, H. D. 1968. "Permanent Farming on the Ukara." In *Smallholder Farming and Smallholder Development in Tanzania.* ed. Hans Ruthenberg. Munich: Weltforum Verlag.

McCown, R. L., G. Haaland, and C. De Haan. 1979. "The Interaction between Cultivation and Livestock Production in Semi-Arid Africa." In *Agriculture in Semi-Arid Environments,* ed. A. E. Hall, G. H. Cannell, and H. W. Lawton. New York: Springer-Verlag.

McC. Netting, Robert. 1968. *Hill Farmers of Nigeria: Cultural Ecology of the Kofyar of the Jos Plateau.* Seattle: University of Washington Press.

McInerney, John P. 1979. *Mechanised Farming in the Central Rainlands of Sudan; Problems and Prospects.* Department of Agricultural Economics and Management, University of Reading, Reading, England.

McInerney, John P., and Graham F. Donaldson. 1975. *The Consequences of Farm Tractors in Pakistan: An Evaluation of IDA Credits for Financing the Mechanization.* Staff Working Paper no. 210. Washington, D.C.: World Bank.

McIntire, John. 1981a. "Rice Production in Mali," In *Rice in West Africa,* ed. Pearson, Stryker, and Humphreys.

———. 1981b. *Two Aspects of Farming in Central Upper Volta: Animal Traction and Intercropping.* ICRISAT West Africa Economics Program, Village Studies Report no. 7. Ouagadougou, Upper Volta: International Crops Research Institute for the Semi-Arid Tropics.

———. 1984. "Policy Environments for Mechanized Agriculture in Sub-Saharan Africa with Special Reference to Animal Traction." ILCA/Systems Research Unit. Addis Ababa, Ethiopia: International Livestock Center for Africa. Processed.

Makonnen, Telahun, and Getachew Assamenew. 1978. *Traditional Agricultural Systems in Ada District, Ethiopia.* ILCA/HL/112/26/78. Addis Ababa: International Livestock Center for Africa.

Manners, R. A. 1962. "Land Use, Trade and the Growth of Market Economy in Kipsigis Country." In *Markets in Africa,* ed. Paul Bohannan and George Dalton. Evanston, Ill.: Northwestern University Press.

Mettrick, H. 1978. "Oxenisation in the Gambia." London: Ministry of Overseas Development. Processed.

Mettrick, H. M., and D. P. James. 1981. *Farm Power in Bangladesh,* vol. 2. Reading, England: University of Reading.

Miracle, Marvin P. 1967. *Agriculture in the Congo Basin.* Madison: University of Wisconsin Press.

Mohammed, Gafar Bashir. 1982. "An Analysis of Smallholder Rainfed Crop Production Systems: A Case Study of the Nuba Mountains Area, Western Sudan." Ph.D. dissertation, Michigan State University, East Lansing.

Moorman, F. R., and D. J. Greenland. 1980. "Major Production Systems Related to Soil Properties in Humid Tropical Africa." In *Priorities for Alleviating Soil-Related Constraints to Food Production in the Tropics.* Los Baños, Laguna, Philippines: International Rice Research Institute.

Morgan, W. B. 1969. "Peasant Agriculture in Tropical Africa." In *Environment and Land Use in Africa,* ed. M. F. Thomas and G. W. Whittington. London: Methuen.

Motilal, G. 1971. "Resource Allocation, Production Efficiency and Comparative Economics on Tractor and Bullock Operated Farms in the Union Territory of Delhi." Ph.D. dissertation, Division of Agricultural Economics, Indian Agricultural Research Institute, New Delhi.

Muckle, T. B. 1971. "Farming in Sierra Leone." *Agriculture* 78(4) 171–74.

Mukwaya, A. B. 1962. "The Marketing of Staple Foods in Kampala, Uganda." In *Markets in Africa,* ed. Paul Bohannan and George Dalton. Evanston, Ill.: Northwestern University Press.

Munzinger, Peter. 1982. *Animal Traction in Africa.* Eschborn: GTZ (German Agency for Technical Cooperation).

Nadjombe, O. 1982. "Taungya Practices in Togo." In *Agro-Forestry in the African Humid Tropics,* ed. L. H. MacDonald. Tokyo: United Nations University.

Narayana, D. L. 1977. *Economics of Tractor Cultivation, Chittoor District, Tirupati.* Tirupati, Andhra Pradesh, India: Sri Venkateswara University.

National Council of Applied Economic Research. 1977. *Demand for Power Tillers.* New Delhi, India.

National Institute of Agricultural Economics (NIAE), Overseas Division. 1981. Evaluation of Tractor Hire Service, Nairobi, Kenya.

———. 1983a. Menufia/Sohag Project, Extracts from visit reports by I. M. Johnson, January and April.

———. 1983b. Engineering and Management Study. Agricultural Mechanisation Services Branch, Ministry of Agriculture, Kenya.

National Research Council. 1982. *Ecological Aspects of Development in the Humid Tropics.* NRC Committee on Selected Biological Problems in the Humid Tropics. Washington, D.C.: National Academy Press.

Neal, M. E. B. 1975. *Farm Mechanisation in the Republic of Sri Lanka.* Rome: FAO.

Nicou, R., and Claude Charreau. 1980. "Mechanical Impedance to Land Preparation as a Constraint to Food Production in the Tropics," In *Soil Related Constraints to Food Production in the Tropics.* Los Baños, Laguna, Philippines: International Rice Research Institute (IRRI).

Nicou, R., and J. L. Chopart. 1979. "Root Growth and Development in Sandy and Sandy Clay Rocks of Senegal." In *Soil Physical Properties and Crop Production in the Tropics,* ed. R. Lal and D. J. Greenland. New York: John Wiley & Sons.

Norman, M. J. T. 1979. *Annual Cropping Systems in the Tropics: An Introduction.* Gainesville: University of Florida Press.

Norman, David W., Emmy B. Simmons, and Henry M. Hays. 1982. *Farming Systems in the Nigerian Savannas.* Boulder, Colorado: Westview Press.

Okawa, Kazushi, M. Shinohara, and M. Umemua. 1966. "Estimates of Long-Term Economic Statistics of Japan since 1868." *Agriculture and Forestry* 9 (Tokyo).

Okigbo, B. N. 1974. "Fitting Research to Farming Systems." Ibadan, Nigeria: International Institute of Tropical Agriculture (IITA). Processed.

———. 1977. "Farming Systems and Soil Erosion in West Africa," In *Soil Conservation and Management in the Humid Tropics,* ed. D. J. Greenland and R. Lal. New York: John Wiley & Sons.

Oluoch-Kosura, Willis A. 1983. "An Economic Analysis of Alternative Methods of Land Preparation in Western Province, Kenya." Ph.D. dissertation, Cornell University, Ithaca, N.Y.

Overseas Development Association (ODA). 1972. Corps of Specialists, Agricultural Machinery Expert on a visit to Malaysia to view the Muda Scheme and look into the plans on Bank Pertanian to enter the Tractor Hire Business. London: ODA.

———. 1973. *The Dwanga Project.* Vols. 11, 17. London: ODA.

———. 1981. Report on the ODA Agricultural Mechanisation Project, Kadugli, South Kordofan, Sudan. London: ODA.

———. 1984. Report on the Nuba Mountains Smallholder Mechanisation Scheme, Sudan. London: ODA.

Owino, F. 1982. "Agro-Forestry Developments in Kenya: Prospects and Problems," In *Agro-Forestry in the African Humid Tropics,* ed. L. H. MacDonald. Tokyo: United Nations University.

Pearson, Scott R., J. Dirck Stryker, and Charles P. Humphreys. 1981. *Rice in West Africa: Policy and Economics.* Stanford, Calif.: Stanford University Press.

Pingali, Prabhu L., and Hans P. Binswanger. 1984. "Population Density and Agricultural Intensification: A Study of the Evolution of Technologies in Tropical Agriculture:" Research Unit Report ARU22. Washington, D.C.: World Bank, Agriculture and Rural Development Department.

Pollard, S. 1983. "The National Farm Power Ownership Study, May 1982." Ministry of Agriculture, AMTU Nakuru, Agricultural Economics Unit. Nakuru, Kenya. Processed.

Pollard, S., A. Metianu, and A. Simmons. 1979. *Evaluation of Tractor-Hire Pools in Swaziland.* London: ODA.

Poulsen, Gunnar. 1978. Man and Tree in Tropical Africa: Three Essays on the Role of Trees in the African Environment. Ottawa: International Development Research Corporation.

Pratt, David J. 1984. "Ecology and Livestock." In *Livestock Development in Sub-Saharan Africa: Constraints, Prospects and Policy,* ed. James R. Simpson and Phylo Evangelo. Boulder, Colorado: Westview Press.

Punjab, Government of (India). 1974. "Economics of Tractor Cultivation and Economics of Production and Cultivation Practices of High Yielding Varieties of Wheat, Maize and Paddy in Punjab, 1969–70 to 1971–72." Combined report.

Purseglove, J. W. 1946. "Land Use in the Over-Populated Areas of Kabale, Kigezi District, Uganda." *East African Agricultural Journal,* October, 74–83.

Raj, K. N. 1973. "Mechanisation of Agriculture in India and Sri Lanka." In *Mechanisation and Employment in Agriculture: Case Studies from Four Continents.* Geneva: International Labour Office.

Ranaivozafy, P. 1978. "Animal Traction in the Malgache Agriculture: The Problems" (in French). Doctoral dissertation, École Nationale Vétérinaire Maison Alfort, Madagascar.

Raulin, H. 1967. *La Dynamique des techniques agraires en Afrique tropicale du Nord.* Paris: Centre National de Recherche Scientifique (CNRS).

Reading, University of. 1978. Bangladesh Rural Mechanisation Study. Progress Reports. Department of Agricultural Economics and Management, G. J. Gill, Field Director.

———. 1981. "Group Farming in North West Nigeria." Department of Agricultural Economics and Management, Development Study no. 18.

Richards, A. I. 1961. "Land, Labour, and Diet in Northern Rhodesia." Oxford: International Institute of African Languages and Culture.

Richards, Alan. 1982. *Egypt's Agricultural Development, 1800–1980: Technical and Social Change.* Boulder, Colo.: Westview Press.

Richards, Alan., and L. P. Martin, eds. 1983. *Migration, Mechanization and Agricultural Labour Markets in Egypt.* Boulder, Colo.: Westview Press.

Roche, L. 1974. "The Practice of Agri-Silviculture in the Tropics, with Special Reference to Nigeria." In *Shifting Cultivation and Soil Conservation in Africa.* FAO Soils Bulletin no. 24. Rome: FAO.

Roesch, M. 1984. *Des Exploitants agricoles face au commerce prive et à* un project de developpement rural: le cas de la region Maradi (Niger). Montpellier: Institut de Recherches Agronomiques Tropicales.

Rotenham, D. von. 1968. "Cotton Farming in Sukumaland." In *Smallholder Farming and Smallholder Development in Tanzania,* ed. Hans Ruthenberg. Munich: Weltforum Verlag.

Rounce, N. V. 1949. *The Agriculture of the Cultivation Steppe.* Cape Town: Longmans, Green and Co.

Ruthenberg, Hans. 1980. *Farming Systems in the Tropics.* 3d ed. Oxford: Clarendon Press.

Ruttan, V. W. 1980. *Institutional Factors Affecting the Generation and Diffusion of Agricultural Technology: Issues, Concepts and Analysis.* Technology and Employment Program, WEP 2–22/WP 67. Geneva: International Labour Office (ILO).

———. 1982. *Agricultural Research Policy.* Minneapolis: University of Minnesota Press.

Saefuddin, Jusef. 1981. "Site Description." Paper presented at a Workshop on the Consequences of Small Rice Farm Mechanization in Asia, 14–18 September. Los Baños, Laguna, Philippines: International Rice Research Institute (IRRI).

Sanchez, Pedro A. 1976. *Properties and Management of Soils in the Tropics.* New York: John Wiley & Sons.

Sapre, S. P. 1969. *A Study of Tractor Cultivation in Shahada.* Gokhale Institute Monograph Series no. 7. Poona, India: Gokhale Institute of Politics and Economics.

Sargent, M. W., J. A. Lichte, P. J. Matlon, and R. Bloom. 1981. *An Assessment of Animal Traction in Francophone West Africa.* African Rural Economy Working Paper no. 34. East Lansing: Michigan State University.

Schapera, I. 1943. *Native Land Tenure in the Bechuanaland Protectorate.* Cape Town: Lovedale Press.

Schneider, Harold K. 1984. "Livestock in African Culture and Society." In *Livestock Development in Sub-Saharan Africa,* ed. James R. Simpson and Phylo Evangelou. Boulder, Colo.: Westview Press.

SEAE (Secrétariat d'État aux Affaires Etrangères Charge de la Cooperation). 1972. *Étude de l'Evolution des Facteurs de Production Mis en Place pendant les 10 Dernières Années, et Leurs Effets.* Republique du Dahomey: CEEMAT.

Seager, P. J., and R. S. Fieldson. 1984. "Public Tractor Hire and Equipment Hire Schemes in Developing Countries (with Special Emphasis on Africa)." Research Unit Report no. ARU 30. Washington, D.C.: Agriculture and Rural Development Department, World Bank.

Sharma, R. K. 1974. *Economics of Tractor Cultivation: A Study in Karnal District, Haryana.* Delhi, India: Agricultural Economics Research Center, University of Delhi.

Sierra Leone, Government of, Ministry of Agriculture and Forestry. 1976. "Mechanical Cultivation Report."

———. 1977a. "Mechanical Cultivations."

———. 1977b. "Tractor Returns."

Simpson, I. G., and M. C. Simpson. 1978. Alternative Strategies for Agricultural Development in the Central Rainlands of the Sudan. Rural Development Study no. 3. Leeds, England: University of Leeds.

Singh, I. J. 1977. "Appropriate Technologies in Tanzanian Agriculture: Some Empirical and Policy Considerations, Annex VII." In *Tanzania Basic Economic Report.* Washington, D.C.: World Bank.

Sleeper, Jonathan A. 1979. *An Economic Analysis of the Role of Ox-Plowing and Cattlefeeding in the Stratification of West African Livestock Production.* Center for Research on Economic Development Working Paper no. 4. Ann Arbor: University of Michigan.

Smeyers, F. 1979. "The Case for the Power Tiller." FAO/World Bank Cooperative Program. Washington, D.C. Processed.

Southall, A. W. 1961. "Population Movements in East Africa," In *Essays on African Population,* ed. K. M. Barbour and R. M. Prothero. London: Routledge & Kegan Paul.

Southworth, H. M. 1972. "Some Dilemmas of Agricultural Mechanization." In *Experience in Farm Mechanization in Southeast Asia,* ed. H. M. Southworth and Milton Barnett. New York: Agricultural Development Council.

Spencer, Dunstan S. C. 1981. "Rice Policy in Sierra Leone." In *Rice in West Africa: Policy and Economics,* ed. Pearson, Stryker, and Humphreys.

Starkey, Paul. 1981. "Farming with Work Oxen in Sierra Leone." Work Oxen Project Report, Ministry of Agriculture and Forestry. Sierra Leone: Freetown.

———. 1982. "N'Dama Cattle as Draught Animals in Sierra Leone." *World Animal Review* (FAO), no. 42:19–26.

Swaziland, Government of, Ministry of Agriculture. 1979. Private Tractor Owner Survey.

———. 1980a. 6th Report on the Operation of Tractor Hire Pools in Rural Development Areas.

———. 1980b. Farm Management Survey. Report no. 3. Mahlangatsha Rural Development Area.

Tanzania, United Republic of. 1983a. "Tractor Rehabilitation Project." Agricultural Planning and Marketing Division. Dar es Salaam: Ministry of Agriculture.

———. 1983b. "The Agricultural Policy of Tanzania." Dar es Salaam: Ministry of Agriculture.

Taylor, H. M. 1980. "Mechanical Impedance to Root Growth." In *Soil Related Constraints to Food Production in the Tropics.* Los Baños, Laguna, Philippines: International Rice Research Institute.

Tchoume, T. 1982. "Agro-Forestry Possibilities in the Oil Palm Plantations in the Ivory Coast." In *Agro-Forestry in the African Humid Tropics,* ed. L. H. MacDonald. Tokyo: United Nations University.

Thapa, Ganesh Bahadjr. 1979. "The Economics of Tractor Ownership and Use in Selected Districts of the Nepal Terai." Master's thesis, University of the Philippines, Los Baños.

Tiffen, Mary. 1976. *The Enterprising Peasant: Economic Development in Gombe Emirate North Eastern State, Nigeria 1900–1968.* London: Her Majesty's Stationery Office.

Tothill, J. D. 1940. *Agriculture in Uganda.* London: Oxford University Press.

———. 1948. *Agriculture in the Sudan.* London: Oxford University Press.

Toulin, Camilla. 1983. "Herders and Farmers or Farmer-Herders and Herder-Farmers?" Pastoral Network Paper. London: Overseas Development Institute.

Trapnell, C. G., and J. N. Clothier. 1937. *The Soils, Vegetation and Agricultural Systems of North Western Rhodesia.* Lusaka: Government of Northern Rhodesia.

Tuluy, A. Hasan. 1981. "Costs and Incentives in Rice Production in Senegal." In *Rice in West Africa,* ed. Pearson, Stryker, and Humphreys.

Uchendu, Victor C., and Kenneth R. M. Anthony. 1975a. *Agricultural Change in Teso District, Uganda.* Nairobi: East African Literature Bureau.

———. 1975b. *Agricultural Change in Kisii District, Kenya.* Nairobi: East African Literature Bureau.

Uganda, Government of, Department of Agriculture Special Development Section. 1967. Annual Report 1967.

———. 1968. Annual Report 1968.

USAID (U.S. Agency for International Development). 1973. *Sudan. Economic and Statistical Data.* Washington, D.C.: USAID.

———. 1979. "ERA 2000. Mechanization of Egyptian Agriculture." Washington, D.C.: USAID. Processed.

Uzureau, C. 1970. *Evolution de l'equipment agricole en Republique Centrafricaine.* Antony: Centre d'Ȇtude et d'Experimentation de Mécanisme Agricole Tropical (CEEMAT).

Wagner, Gunter. 1949. *The Bantu of North Kavirondo.* Vol. 1. London: Oxford University Press.

Walker, T. S., and K. G. Kshirsagar. 1981. *The Village Level Impact of Machine Threshing and Implications for Technology Development in Semi-Arid Tropical India.* ICRISAT Economics Program, Progress Report no. 27. Hyderabad: International Crops Research Institute for the Semi-Arid Tropics.

WARDA (West Africa Rice Development Association). 1977. "Large Scale Mechanized Rice Production in Northern Ghana." Case Study no. 3. Monrovia, Liberia: WARDA.

Webster, C. C., and P. N. Wilson. 1980. *Agriculture in the Tropics.* 2d ed. London: Longman.

Whitney, Thomas R. 1981. "Changing Patterns of Labor Utilization, Productivity, and Income: The Effects of Draft Animal Technology on Small Farms in Southeastern Mali." M.S. thesis, Purdue University.

Index